T0295426

NON-ISOLATED DC-DC CONVERTERS FOR RENEWABLE ENERGY APPLICATIONS

NON-ISOLATED DC-DC CONVERTERS FOR RENEWABLE ENERGY APPLICATIONS

Frede Blaabjerg, Mahajan Sagar Bhaskar, and Sanjeevikumar Padmanaban

CRC Press
Taylor & Francis Group
Boca Raton London New York

CRC Press is an imprint of the
Taylor & Francis Group, an **informa** business

First edition published 2021 by
CRC Press
6000 Broken Sound Parkway NW, Suite 300, Boca Raton, FL 33487-2742

and by CRC Press
2 Park Square, Milton Park, Abingdon, Oxon, OX14 4RN

© 2021 Frede Blaabjerg, Mahajan Sagar Bhaskar, Sanjeevikumar Padmanaban

CRC Press is an imprint of Taylor & Francis Group, LLC

ISBN: 978-0-367-65458-0 (hbk)
ISBN: 978-0-367-65460-3 (pbk)
ISBN: 978-1-003-12953-0 (ebk)

Typeset in Times New Roman
by MPS Limited, Dehradun

Contents

Preface

Renewable energy sources (RES) are becoming more versatile day by day due to government support and a carbon dioxide (CO_2) emission-reduction policy. For reducing the greenhouse gas emissions as of EU 2030 and 2050 policies, a joint effort of all corners of the globe. In such circumstances, photovoltaic (PV) energy generation is an excellent example of green energy generation through the various parallel arrangements of small voltage-generating cells or modules. This is the direct use of synchronous generators to transfer power to the primary grid from hydroenergy plants, geothermal energy plants, and biofuel energy plants. However, the photovoltaic energy-generation systems require the power electronic converters system to satisfy the demand for real-time applications or electric grids. Therefore, for real-time applications or before feeding energy to the grid via an inverter, photovoltaic systems linked with DC-DC converters have high-voltage conversion ratio capability. Henceforth, a DC-DC power converter is a vital constituent in the photovoltaic power conversion stage.

In this book, we have focused on the presentation of the developments, a new series of power electronics DC-DC converter in an extra dimension, for trends setting future demands. We have presented the work with basic circuitry design based on circuitry laws, followed by numerical modeling and control, and further hardware implementation for proof concepts. This book also provides intensive investigation studies:

- Usually for a new series of non-isolated (unidirectional) DC-DC converter configurations for RES applications. The authors have presented a comprehensive state-of-the-art review of various non-isolated DC-DC power converters that illustrate the capability to convert the low voltage into a high voltage; thus, not suitable for PV energy applications.
- The authors have taken extensive measures and careful governance with the design of new DC-DC multistage power converter topologies to ensure a high-voltage conversion ratio by employing a new circuitry placement of reactive elements and/or semiconductor devices (MosFET).
- We have shown a new breed of DC-DC multistage power converters called the "X-Y converter family," suggested for PV applications and configuration by utilizing switched-inductor, switched-capacitor, voltage-lift switch-capacitor, and voltage-lift switched-capacitor; voltage doubler; and multiplier boosting techniques, etc.
- We have shown the development of novel, non-isolated DC-DC converters (unidirectional) based on the Z-source and Quasi Z-source-based configurations.
- An original Transformer and Switched-Capacitor (T-SC) based multistage power converter is also suggested for high-voltage/low-current PV applications. The configuration is based on the benefit of combination with the boost converter, T-SC.

- New Nx IMBC (Nx Interleaved Multilevel Boost Converter) or Cockcroft Walton (CW) Voltage Multiplier based Multistage/Multilevel Power Converter (CW-VM-MPC) converter topologies—purpose is to achieve a maximum voltage conversion ratio by utilizing the features of the Cockcroft Walton (CW) voltage multiplier.

For all developments, we have given priority to the theoretical background behind the voltage transfer gain ratio: the advantage of each converter. Further, we have suggested the multistage power converter suitably fits the PV application, and we compared the solution with other recent multistage power converters in terms of voltage conversion ratio, number of devices, and costs. For initial assessment with proof of concepts, we have designed the converter based on the power requirement and components, and tested and verified using numerical simulation software and prototype hardware test bench on a laboratory scale. We elaborated the hardware prototype implementation for DC-DC multistage power converters without magnetic components, T-SC based multistage power converters, and Nx-IMBC configurations. All results presented were based on experimental and numerical simulation tasks in good agreement with original theoretical hypotheses developed.

We believe this book provides essential support and guidelines for a designed, systematic approach for numerical modeling—validation for new trends in dimensions of DC-DC high-voltage generation converter—and selection based on applications.

Warm regards,
Frede Blaabjerg
Mahajan Sagar Bhaskar
P. Sanjeevikumar

Keywords

Power Electronics, Non-Isolated, Unidirectional, Multistage DC-DC Converter, High-voltage Conversion Ratio, Boost Converter, High Output Voltage, Maximum Power Point Tracking (MPPT), Switched-Capacitor, Switched-Inductor, Diode-Capacitor Circuit, X-Y Converter, Z-source, Quasi Z-source, Magnetic Free, T-SC Configuration, Multilevel DC-DC Converter, Cockcroft Walton (CW) Voltage Multiplier, Photovoltaic (PV), Renewable Energy Systems (RES)

Acknowledgments

We wish to express our gratitude to the Center of Reliable Power Electronics (CORPE), Aalborg University, Department of Energy Technology, Aalborg, Denmark; Renewable Energy Lab (REL), College of Engineering, Prince Sultan University, Riyadh, Saudi Arabia; and the Center for Bioenergy and Green Engineering, Department of Energy Technology, Aalborg University, Esbjerg, Denmark for supporting us with various resources and time to bring this long, extensive original research work into a successful book.

We are thankful to all internal/external colleagues, international collaborates, family members, and friends who have supported us directly/indirectly through several means.

Thanks to one and all!

About the Authors

Frede Blaabjerg was with ABB-Scandia, Randers, Denmark, from 1987 to 1988. From 1988 to 1992, he earned a Ph.D. degree in electrical engineering from Aalborg University in 1995. He became an assistant professor in 1992, an associate professor in 1996, and a full professor of power electronics and drives in 1998. In 2017, he became a villum investigator. He is honoris causa at University Politehnica Timisoara (UPT), Romania and Tallinn Technical University (TTU) in Estonia. His current research interests include power electronics and its applications such as in wind turbines, PV systems, reliability, harmonics, and adjustable speed drives. He has published more than 600 journal papers in the fields of power electronics and its applications. He is the co-author of four monographs and editor of ten books in power electronics and its applications. He has received 32 IEEE Prize Paper Awards, the IEEE PELS Distinguished Service Award in 2009, the EPE-PEMC Council Award in 2010, the IEEE William E. Newell Power Electronics Award 2014, the Villum Kann Rasmussen Research Award 2014, the Global Energy Prize in 2019, and the 2020 IEEE Edison Medal. He was the editor-in-chief of the IEEE Transactions on Power Electronics from 2006 to 2012. He has been a distinguished lecturer for the IEEE Power Electronics Society from 2005 to 2007 and for the IEEE Industry Applications Society from 2010 to 2011 as well as 2017 to 2018. From 2019 to 2020 he served as president of the IEEE Power Electronics Society. He is vice president of the Danish Academy of Technical Sciences, too. He was nominated in 2014–2019 by Thomson Reuters as one of the 250 most-cited researchers in engineering in the world. In 2017, he became honoris causa at University Politehnica Timisoara (UPT), Romania.

Mahajan Sagar Bhaskar received a bachelor's degree in electronics and telecommunication engineering from the University of Mumbai, Mumbai, India in 2011; a master's degree in power electronics and drives from the Vellore Institute of Technology, VIT University, India in 2014; and Ph.D. in electrical and electronic engineering, University of Johannesburg, South Africa in 2019. Currently, he is with Renewable Energy Lab, Department of Communications and Networks Engineering, College of Engineering, Prince Sultan University, Riyadh, Saudi Arabia. He has published scientific papers in the field of power electronics, with particular references to XY converter family, multilevel DC/DC and DC/AC converters, and high-gain converters. Dr. Mahajan has authored more than 100 scientific papers and has received the Best Paper Research Paper Awards from IEEE-CENCON'19, IEEE-ICCPCT'14, IET-CEAT'16, and ETAEERE'16 sponsored lecture note in electrical engineering, Springer book series. He is a senior member of IEEE, IEEE Industrial Electronics,

Power Electronics, Industrial Application, and Power and Energy, Robotics and Automation, Vehicular Technology Societies, Young Professionals, and various IEEE Councils and Technical Communities. He is a reviewer member of various international journals and conferences including IEEE and IET. He received the IEEE ACCESS award "Reviewer of Month" in January 2019 for his valuable and thorough feedback on manuscripts, and for his quick turnaround on reviews. He is an associate editor for IET Power Electronics.

Sanjeevikumar Padmanaban (M'12–SM'15) received a bachelor's degree in electrical engineering from the University of Madras, Chennai, India, in 2002; a master's degree (Hons.) in electrical engineering from Pondicherry University, Puducherry, India, in 2006; and a Ph.D. degree in electrical engineering from the University of Bologna, Bologna, Italy, in 2012. He was an associate professor with VIT University from 2012 to 2013. In 2013, he joined the National Institute of Technology, India, as a faculty member. In 2014, he was invited as a visiting researcher at the Department of Electrical Engineering, Qatar University, Doha, Qatar, funded by the Qatar National Research Foundation (Government of Qatar). He continued his research activities with the Dublin Institute of Technology, Dublin, Ireland, in 2014. Further, he served an associate professor with the Department of Electrical and Electronics Engineering, University of Johannesburg, Johannesburg, South Africa, from 2016 to 2018. Since 2018, he has been a faculty member with the Department of Energy Technology, Aalborg University, Esbjerg, Denmark. He has authored more than 300 scientific papers.

Dr. Sanjeevikumar was the recipient of the Best Paper with Most Excellence Research Paper Award from IET-SEISCON'13, IET-CEAT'16, IEEE-EECSI'19, IEEE-CENCON'19 and five best paper awards from ETAEERE'16 sponsored Lecture Notes in electrical engineering, Springer book. He is a fellow of the Institution of Engineers, India, the Institution of Electronics and Telecommunication Engineers, India, and the Institution of Engineering and Technology, U.K. He is an editor/associate editor/editorial board for refereed journals, in particular the IEEE SYSTEMS JOURNAL, IEEE Transaction on Industry Applications, IEEE ACCESS, IET Power Electronics, IET Electronics Letters, and Wiley-International Transactions on Electrical Energy Systems, Subject Editorial Board Member—Energy Sources—Energies Journal, MDPI, and the subject editor for the IET Renewable Power Generation, IET Generation, Transmission and Distribution, and FACTS journal (Canada).

1 Introduction

1.1 MOTIVATION AND RESEARCH FORMULATION

Nowadays, fossil fuel reserves are dwindling because of excessive burning and utilization of fossil fuels for industrial, domestic, transportation, and other applications, which has led to the emission of greenhouse gases (GHGs) [1–5]. The average temperature of the earth is rising slowly because of the emission of GHGs. According to a report presented by the International Energy Agency (IEA) in 2016, 37% of the Total Energy Consumption (TEC) of the world is from petroleum, 29% from natural gases, 15% from coal, 9% from nuclear, and 10% via renewable energy sources [6,7]. The report notably pointed out that the hotness of the surface of the earth could rise as early as 2050 if no scheme taken to control the current emission rate of GHG. In last few years, to preserve the clean environment and to reduce the emission rate of GHG, many countries have adopted various government and private schemes and policies (e.g., the energy taxes imposed on the fuel regulatory [8,9]). This is the intrinsic motivation of many researchers throughout the world to find new renewable energy sources and extensive measure through strategies to reduce CO_2 emission by 70% in 2050 as the target goal. The IEA anticipates that the developing countries will elevate their energy consumption more rapidly than developed ones and, to fulfill the energy requirement, will nearly double their presently installed power generation facility by the year 2020. Figure 1.1 depicts the TEC of the world from the year 1980 to 2030 [10–13].

According to the report stated by the IEA, the feasibility of an electric grid to remote areas (rural locations) in developing countries are limited, over 1.3 billion people are living without access to electricity [14,15]. Henceforth, there is a requirement of low-cost, renewable energy from local availability and resource to reduce carbon emission. Global cumulative photovoltaic (PV) energy capacity is depicted by Figure 1.2 (1996–2020) and estimation (E) by 2020 [10,16,17].

PV technology is a fast-growing sector among renewable energy systems because of reliable, promising, and favorable sources, with advantages such as pollution-free, long life, low maintenance, etc. [18–20] for real-time applications or before feeding energy to the grid via photovoltaic inverter systems linked with a DC-DC converter, which has high-voltage conversion ratio capabilities. Thus, DC-DC power converter configurations are the major constituent in the photovoltaic power conversion stage [21–24].

Recently, in power-electronics-based research articles, many DC-DC converter configurations addressed various photovoltaic electric drive applications with utilizing a switched capacitor, switched inductor, voltage multiplier, Z source and Quasi Z source, and transformers, etc. reactive networks [25,26]. However, to enable different applications, design of a suitable power converter is a challenging

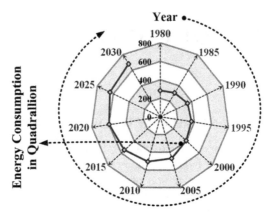

FIGURE 1.1 Total Energy Consumption of the World from the Year 1980 to 2030 [10].

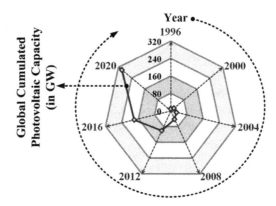

FIGURE 1.2 Global Cumulated Photovoltaic Energy Capacity from the Year 1996–2012 and Estimation by 2020 [10].

issue for the fluctuating nature of photovoltaic energy sources, losses, and weight of converters. DC-DC power converters are more necessary in order to design low-cost, high-power, and efficient photovoltaic electric drives system. Increasing the voltage conversion ratio and utilizing few components with low voltage stress becomes a dominant factor in order to have additional freedom. Non-isolated multistage DC-DC converters widely replace conventional DC-DC converters and isolated converters by their performance in higher conversion ratios, high power, low volume, low weight, and low voltage stresses across switches [27–31]. Combining multiple boosting circuits with conventional DC-DC boost converters could solve the high-voltage conversion ratio for photovoltaic energy applications [32–35]. This book discusses a non-isolated DC-DC multistage power converter configuration based on the new arrangement of reactive elements and semiconductor devices in response to the demand for high-voltage conversion ratio DC-DC power converters. The suggested multistage converter configurations provide a robust optimum solution to achieve high-voltage conversion ratios for photovoltaic energy applications.

1.2   BOOK-FOCUSED AIM

Our primary aim is to present this book with the design and development of non-isolated DC-DC (unidirectional) multistage power converter configurations for renewable applications. All verifications of functionality, mode of operation tested, and proof of concept are through numerical simulation software and hardware lab prototype implementation for applications.

This book is focused on:

1. Presenting state-of-the-art PV-based power electronics DC-DC converters and maximum power point tracking (MPPT) algorithms, which provide a viable solution in terms reduced cost, volume, and weight.
2. The extensive role of the characteristics, advantages, and disadvantages of the existing non-isolated multistage DC-DC power converters.
3. Design and development of new non-isolated multistage DC-DC power converter configurations to produce a high-voltage conversion ratio and a minimal number of reactive elements/semiconductor switches (MOSFET) used.
4. Design and development of a new breed of DC-DC power converters named X-Y converter families, based on the voltage-lift technique.
5. Hardware lab-prototype development and proof of validation shown for
   • the multistage power converter without magnetic component
   • T-SC multistage DC-DC power converter, multistage power converter, based on the Cockcroft Walton (CW) voltage multiplier.

We author kept our goal to provide new generation power electronics converters in particular to DC-DC circuits topologies with low-cost, minimal passive/active components with reliable and durable solutions. Particular, the application found suits the renewable energy application (PV solar, Wind, Fuel cell, etc.) and wherever DC high voltage rating for power requirement.

Brief outlines of chapters:

Chapter 2: "Power Electronics Photovoltaic Configurations and MPPT Algorithms"

Survey-based on the Central Inverter Photovoltaic Configuration (CI-PVC)—String Inverter Photovoltaic Configuration (SI-PVC)—AC Module Photovoltaic Configuration (ACM-PVC)—Multi-String Inverter Photovoltaic Configuration (MSI-PVC)—PV grid-connected power systems. We discuss the advantages and drawbacks of all the existing PV systems in details—survey on MPPT algorithms used in PV applications—merit and demerit of each MPPT algorithm and conclusion with the review.

Chapter 3: "Non-isolated Unidirectional Multistage DC-DC Power Converter Configurations"

State-of-the-art review of non-isolated (unidirectional) multistage DC-DC power converters. Classification of DC-DC and multistage DC-DC power converter configurations based on the boosting techniques, including advantages, dis-advantages, and conspicuous features of each configuration. The comparison is based on the number of components and stages, in terms of efficiency, cost, re-liability, and applications.

Chapter 4: "X-Y Power Converter Family: A Breed of DC-DC Multistage Configurations for Photovoltaic Applications"

This chapter deals with a new breed of multistage power converters called the X-Y power converter family, designed based on the arrangement of reactive elements and semiconductor devices and classified into three categories: two-stage X-Y, three-stage X-Y, and N-stage X-Y power converter configurations. Analysis, working modes and validation through numerical simulation software are presented. Finally, we have given the specific practical conclusions for a better selection of converters for application by the readers.

Chapter 5: "Self-Balanced DC-DC Multistage Power Configuration without Magnetic Components for Photovoltaic Applications"

Self-balanced DC-DC multistage power configurations without a magnetic component is discussed with the MPPT algorithm, working modes, and equivalent ON state and OFF state circuit. Loss analysis of the semiconductor device, and its effects on voltage conversion ratio, the complete theoretical background and design equation, and rating of components is illustrated; configuration comparison in terms of the number of semiconductor devices, capacitors, and voltage con-version ratio, experimental prototype and numerical simulation results for the multistage power converters is presented. We have provided the practicability based on test results observed as a conclusion for converter selection for appli-cation in real-time.

Chapter 6: "T-SC MPC: Transformer and Switched-Capacitor Based DC-DC Multistage Power Converter for Photovoltaic Applications"

Transformer and Switched-Capacitor based DC-DC Multistage Power Converter (T-SC MPC) design and development are presented. We discuss multistage con-figurations advantages and disadvantages in terms of the T-SC MPC and MPPT algorithm. Also discussed is the original theoretical background, analysis, design and working modes, equivalent ON state and OFF state circuit, steady-state ana-lysis, and comparison based on the cost and the number of components is elabo-rated, along with experimental prototypes and numerical simulations results provided with practical feasibility conclusions.

Chapter 7: "Cockcroft Walton Voltage Multiplier based Multistage/ Multilevel Power Converter Configurations for Photovoltaic Applications"

This chapter deals with a CW voltage multiplier-based multistage (here also called multilevel) power converter configurations. We discuss comparative studies with existing power converter configurations. We suggest Nx IMBC multilevel power converters and utilizing the interleaved concept. The complete theoretical background, analysis, design, and working modes are presented. Comparison in terms of cost components and voltage conversion gain ratio, the effect of the in-ternal series resistance of an inductor, etc. are elaborated. We present numerical simulations and experimental prototype results for interleaved Nx converter con-figurations with a conclusion on practicability.

Chapter 8: "Conclusion and Future Direction"
We have given overall conclusions of this book based on the state-of-the-art review, design, and development, testing using numerical simulation, and experimental investigation for power converter configurations. Finally, we have given a set of suggestions for future direction and possible extended research activities.

References
An extensive survey collection of articles is presented, and it could provide guidelines for the reader to gain more insight on the technical aspects of converters.

2 Power Electronics PV Configurations and Maximum Power Point Tracking Algorithms

2.1 INTRODUCTION

In the present and future, many applications use and will use electrical energy; thus, the demand for electrical energy consumption is increasing. It is expected that 60–70% of total energy consumption will be transferred into electrical energy [36–38]. We need to find a solution to generate, transfer, and distribute electrical energy with maximum effectiveness due to emerging climate changes. Thus, the main issue for the present is appropriate consumption and utilization of energy sources. With the rising demand of renewable energy, PV energy resources are becoming more accepted due to cost, cleanliness, limitless energy compared to other sources [39–42]. PV energy cost is declining with continuous improvement in power and semiconductor technology sectors.

Power electronics play a vital role in the conversion of PV energy to electrical energy. In recent years, DC-DC and DC-AC power converter technologies are have been a part of remarkable advancements in developing PV industries [43,44]. In the next section, we discuss PV configuration systems. This chapter also deals with maximum power point tracking (MPPT) algorithms used in PV applications to extract maximum power from the PV source. We will discuss the merits and de-merits of each MPPT algorithm in detail.

2.2 POWER ELECTRONICS PV CONFIGURATIONS

There is continuous improvement in the DC-DC and DC-AC power converter configurations to reduce the cost of PV systems with higher efficiency. Based on the power converters' arrangement and existing structures, PV systems are divided into the following four main configurations:

- Central inverter PV configuration (CI-PVC)
- String inverter PV configuration (SI-PVC)
- AC-module PV configuration (ACM-PVC)
- Multi-string inverter PV configuration (MSI-PVC)

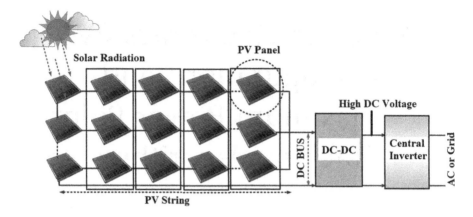

FIGURE 2.1   Central Inverter PV Configuration (CI-PVC).

2.2.1   CENTRAL INVERTER PV CONFIGURATION (CI-PVC)

The central inverter PV configuration (CI-PVC) is designed to transfer generated PV energy to the electrical grid by employing many series and parallel connect PV modules or panels with a central inverter [45–49].

A large number of PV panels or modules are connected in series to form a sufficient power PV string. Large power strings are arranged in parallel and further connected to the central inverter via a DC-DC power converter. Figure 2.1 depicts a general block diagram of the CI-PVC [49]. Initially, inverters with line commutation are used as a central inverter. Inverters with force commutation of Insulated Gate Bipolar Transistors (IGBTs) have lower costs and higher efficiency. Thus, inverters with line commutations are gradually replaced by these inverters. This configuration still has the following demerits:

- It is highly rated and requires many power cables, which decreases the efficiency and increases the implementation cost of the configuration.
- Centralized MPPT is adopted, which leads to more losses; hence, efficiency is low.
- More power losses occur due to unequal characteristics of modules.
- More power losses occur due to string diodes.
- It is not suitable for mass production due to its inflexible design.
- Reliability of the configuration is dependent on the centralized single inverter, and the configuration generates a poor-quality waveform with large current harmonics.

2.2.2   STRING INVERTER PV CONFIGURATION (SI-PVC)

The string inverter PV configuration (SI-PVC) is designed to transfer generated PV energy to the electrical grid by employing a separate inverter for PV strings [45–49]. Many PV panels or modules are connected in series to form the PV string.

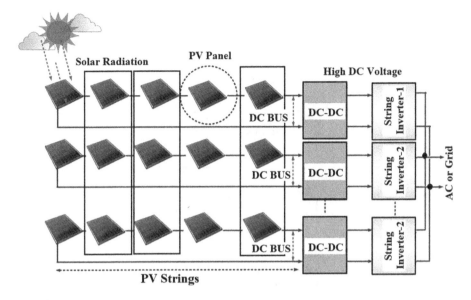

FIGURE 2.2 String Inverter PV Configuration (SI-PVC).

All the PV strings are connected to separate inverters via a separate DC-DC converter. It feeds the outputs of all inverters parallel to the electric grid. No DC-DC boost converter is required if the string voltage is large enough. However, the DC-DC converter is compulsory if fewer numbers of panels are used to form a PV string. Figure 2.2 depicts a general block diagram of the SI-PVC [49]. This configuration has minimum power loss in the "string" diode because a separate DC-DC converter and MPPT is used in this configuration. Therefore, the configuration has a higher efficiency compared to a CI-PVC.

Lower costs are possible due to the mass production of energy when using this configuration. This configuration still has the following demerits:

- It requires several high-rated power cables, which decreases the efficiency and increases the implementation cost of the configuration.
- It requires many power converters to feed PV energy to the grid.
- It requires separate MPPTs for each DC-DC converter.
- It requires complex control circuitry to control the power converter.
- Initial installation cost is higher because of several converters and complex controls.

2.2.3 AC-MODULE PV CONFIGURATION (ACM-PVC)

The AC-module inverter PV configuration (ACM-PVC) is designed by utilizing an AC module inverter between the PV source and electric grid for each PV panel or module [45–49]. Figure 2.3 depicts a general block diagram of an ACM-PVC [49]. This configuration eliminates the mismatch loss, which occurs in unequal

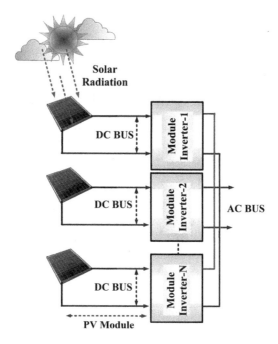

FIGURE 2.3   AC-Module Inverter PV Configuration (ACM-PVC).

characteristics of PV panels. It also minimizes losses because each panel has its own power converter and MPPT.

Due to modularity and simple modifications, this configuration has a plug-and-play quality. Compared to CI-PVC and SI-PVC, the power extraction is improved, and there is no requirement of strings. This configuration still has the following demerits:

- It requires several module inverters to transfer PV energy to the electric grid.
- The configuration has a higher cost for the AC-module inverters.
- The efficiency is not high for high-voltage conversion.
- Synchronization is difficult at the output of the AC-module inverters.

2.2.4   MULTI-STRING INVERTER PV CONFIGURATION (MSI-PVC)

The multi-string inverter PV configuration (MSI-PVC) is designed to transfer generated PV energy to the electrical grid by employing a separate DC-DC power converter for each PV string and common inverter. Many PV panels or modules are connected in series to form a PV string [45–49]. Figure 2.4 depicts a general block diagram of the MSI-PVC [49]. All the PV strings are connected to separate DC-DC converters from boosting the voltage and all the outputs of DC-DC converters are fed to electric grids via a typical central inverter. This configuration is a further development of SI-PVC and CI-PVC. The advantage over CI-PVC is that in MSI-PVC, each PV string is controlled separately through the DC-DC converter. This

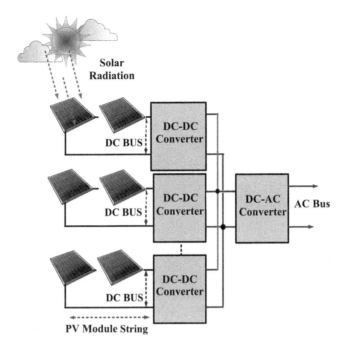

FIGURE 2.4   Multi-String Inverter PV Configuration (MSI-PVC).

configuration has a flexible solution and higher overall efficiency compared to CI-PVC because of the separate MPPT control. This configuration has a higher efficiency compared to SI-PVC because of fewer power converters. This configuration still has the following demerits:

- The cost is higher compared to a CI-PVC for many DC-DC converters.
- Separate MPPT control is required, which makes the control circuit complex.
- Voltage unbalancing occurs at the input side of the inverter.

2.2.5   COMPARISON OF CI-PVC, SI-PVC, ACM-PVC, AND MSI-PVC

In this section, the previously discussed configurations are compared in terms of efficiency, effectiveness related to MMPT, reliability, power range, scalability, manufacturability, and installation cost. The detailed comparison is provided in Table 2.1.

2.3   MAXIMUM POWER POINT TRACKING

PV characteristics are nonlinear for continuous changes in PV irradiation and temperature. MPPT is essential and necessary to extract maximum power from the PV panel. The MPPT concept is based on the maximum power transfer theorem and maximum power will be extracted when the impedance of the network matched with load impedance [50–56]. Maximum power is extracted by varying the duty

TABLE 2.1

Detailed Comparison of CI-PVC, SI-PVC, ACM-PVC, and MSI-PVC

Characteristics	CI-PVC	SI-PVC	ACM-PVC	MSI-PVC
Overall Efficiency	Medium	Very High	High	Medium
Effectiveness Related to MPPT	Low	Medium	High	Medium
Reliability	Very high	High	Low	Medium
Power Range	High	Medium	Low	Medium
Scalability	Very Low	Medium	High	Medium
Manufacturability (Bulkiness and Production)	Low	Medium	High	Medium
Cost	Low	Medium	High	Medium

cycle of a power converter appropriately to match load impedance and source impedance. Figure 2.5 depicts the I-V characteristics of the PV cell for different irradiations. The current increases when irradiation increases. Figure 2.6 depicts the P-V characteristics of the PV cell for different irradiations. The MPP shifts upwards with increasing irradiation.

The I-V and P-V characteristics vary with irradiation and temperature. Therefore, it is not a viable solution to fix the duty cycle of a power converter under such dynamically varying operating conditions. MPPT algorithms are used to track coordinates of MPP (V_{mpp}, P_{mpp}) on P-V characteristics where maximum power is extracted from the source, irrespective of temperatures and radiations. In this section, the following four algorithms are commonly used to track MPP:

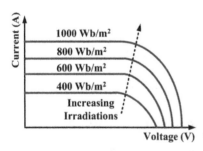

FIGURE 2.5 I-V Characteristics of a PV Cell Under Different Irradiations.

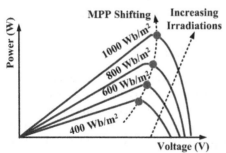

FIGURE 2.6 P-V Characteristics of a PV Cell Under Different Irradiations.

- Perturb and Observe Algorithm
- Incremental Conductance (IC) Algorithm
- Current Sweep Algorithm
- Constant Voltage Algorithm

2.3.1 PERTURB AND OBSERVE ALGORITHM

The perturb and observe algorithm of the MPPT is based on the sampling of the PV voltage, current, and computation of the change in power [57–59].

In this algorithm, the tracker works by continuously changing the PV array voltage. Succeeding perturbation is generated in the required direction if there is an increase or decrease in the output PV power. Based on the succeeding perturbation, the duty cycle of the DC-DC converter is changed, and the process continues to find the MPP. The classical perturb and observe is shown in the algorithm in Figure 2.7. This method also called a hill-climbing algorithm because the algorithm is dependent on the rise and fall of the power curve. The algorithm is simple and with less computation, but has the following demerits:

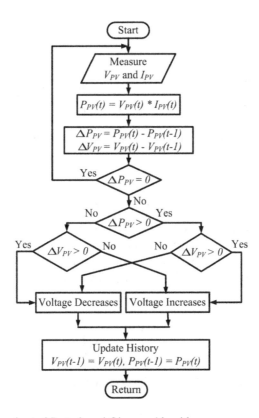

FIGURE 2.7   Flowchart of Perturb and Observe Algorithm.

- Energy loss occurs during oscillation of the operating point around the MPP.
- The controlability decreases under a drastic change in environmental conditions, and the operating point is moving away from the MPP.

2.3.2 INCREMENTAL CONDUCTANCE ALGORITHM

The incremental conductance (IC) MPPT algorithm is based on the measurement of change in the PV voltage and current to predict the effect of voltage change [57,59–61]. The algorithms require more computational controllers, but track the MPP faster compared to the perturb and observe algorithm. This method is based on the concept of slope of the PV curve and uses the change in the ratio of current and voltage $(\Delta I_{PV}/\Delta V_{PV})$ to calculate the signature of the change in the PV power regarding the change in the ratio of power and voltage $(\Delta P_{PV}/\Delta V_{PV})$. This algorithm has the following demerits:

- It needs more computations compared to the perturb and observe algorithm.
- It takes more time to track the MPP compared to the perturb and observe algorithm.

The following numerical (2.1) are used to track the MPP and incremental conductance (IC) MPPT algorithm shown in Figure 2.8.

$$
\left.
\begin{array}{l}
\frac{\Delta P_{PV}}{\Delta V_{PV}} < 0,\ \frac{\Delta I_{PV}}{\Delta V} < -\frac{\Delta I_{PV}}{\Delta V},\ \text{Right side of MPP} \\[2mm]
\frac{\Delta P_{PV}}{\Delta V_{PV}} = 0,\ \frac{\Delta I_{PV}}{\Delta V_{PV}} = -\frac{\Delta I_{PV}}{\Delta V_{PV}},\ \text{At MPP} \\[2mm]
\frac{\Delta P_{PV}}{\Delta V_{PV}} > 0,\ \frac{\Delta I_{PV}}{\Delta V_{PV}} > -\frac{\Delta I_{PV}}{\Delta V_{PV}},\ \text{Left side of MPP}
\end{array}
\right\}
\tag{2.1}
$$

2.3.3 OPEN CIRCUIT CONSTANT VOLTAGE (OCCV) FRACTIONAL ALGORITHM

The OCCV fractional algorithm uses the value of the open-circuit voltage to track the MPP. The open-circuit voltage with zero current is measured to locate the MPP [57,62–64]. The controller uses the fixed ratio (A) between 0.72 and 0.78. Once the controller fixes the ratio (A), and V_{OC}, V_{MPP} is calculated in the following equation (2.2):

$$
\left.
\begin{array}{l}
V_{MPP} = A * V_{OC} \\
0.70 <= A <= 0.78
\end{array}
\right\}
\tag{2.2}
$$

The OCCV fractional algorithm has fewer computations, but has the following demerits:

- There is less efficiency to track the MPP for an approximate fixed ratio of the voltage at the MPP and open-circuit voltage (V_{MPP}/V_{OC}).

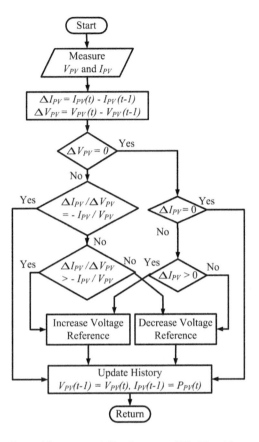

FIGURE 2.8 Flowchart of Incremental Conductance (IC) Algorithm.

- Periodic closing of the power converter is required to calculate the open-circuit voltage.
- A pilot cell is used and carefully chosen to prevent a temporary loss problem and to understand the characteristic of the PV array or panel.
- It is based on the approximation and open-circuit voltage.

2.3.4 SHORT CIRCUIT CURRENT (SCC) FRACTIONAL ALGORITHM

The short circuit current (SCC) fractional algorithm uses the value of the SCC to track the MPP. The SCC is measured to locate the MPP [57,62–64]. The controller uses the fixed ratio (A) between 0.75 and 0.92. Once the controller fixes the ratio (A), and I_{SC}, I_{MPP} is calculated by the following equation:

$$\left.\begin{array}{l} I_{MPP} = A \times I_{SC} \\ 0.78 <= A <= 0.92 \end{array}\right\} \tag{2.3}$$

TABLE 2.2

Comparison of MPPT Algorithms

MPPT Algorithm	Speed	Complexity in Implementation	Periodic Tuning	Sensing Parameter
Perturb and Observe	Varies	Low	No	Voltage and Power
Incremental Conductance	Varies	Medium	No	Voltage, Current, Power
Open Circuit Constant Voltage	Higher	Low	Yes	Voltage
Short Circuit Current	Higher	Medium	Yes	Current

The SCC fractional algorithm has fewer computations, but has the following demerits:

- Efficiency to tracking MPP is lesser because of an approximated fixed ratio of current at MPP and short circuit current (I_{MPP}/I_{SC}).
- It requires the additional power switch to measure SCC.
- It is based on the approximation and open-circuit voltage.

2.3.5 COMPARISON OF MPPT ALGORITHMS

In this section, the previously discussed MPPT algorithms are compared in terms of speed, periodic tuning, complexity in implementation, and sensing parameter—the comparison of the MPPT algorithms are provided in Table 2.2.

2.4 CONCLUSION

A detailed discussion on the PV power converter system was presented. The configurations of the PV power converter system were classified into four main configurations: 1) central inverter PV configuration (CI-PVC), 2) string inverter PV configuration (SI-PVC), 3) AC-module PV configuration (ACM-PVC), and 4) multi-string inverter PV configuration (MSI-PVC). We discussed the drawbacks of all PV configurations in detail. A detailed discussion on MPPT algorithms was also presented. Four popular algorithms, including 1) perturb and observe, 2) incremental conductance, 3) open circuit constant voltage (OCCV) fractional algorithm, and 4) short circuit current (SCC) fractional algorithm, were discussed and compared in terms of speed, complexity in implementation, periodic tuning, and sensing parameters. Based on the review, it was concluded that the PV system is influenced by the chosen power electronics converter configurations and its MPPT control strategy. The PV energy will play a significant role in future power systems enabled by power technology.

3 Non-Isolated Unidirectional Multistage DC-DC Power Converter Configurations

3.1 INTRODUCTION

In the last decades, numerous DC-DC multistage converter configurations are proposed and implemented for the vehicular power train, renewable energy, and electric drives applications. By factors of efficiency, cost, and size, the DC-DC multistage converter configurations were shown good suitability of adaption with photovoltaic power applications [65–67]. The choice of the appropriate DC-DC multistage converter configurations is essential to plan a photovoltaic system with higher efficiency and lower cost [68–70]. Thus, crucial multistage converter configurations are reviewed from the recently addressed converters and state-of-the-art of non-isolated unidirectional multistage converter configurations presented in this chapter. The conspicuous features of all the presented topologies are discussed.

3.2 DC-DC POWER CONVERTER CONFIGURATIONS

The DC-DC power converter configurations are shown in Figure 3.1 [71]. A DC-DC power converters, AC-AC power converters, AC-DC power converters, and DC-AC power converters are the general configurations of power electronics converters. The DC-DC and DC-AC power converters are significant configurations of power electronics converters and play a vital role in photovoltaic and industrial applications [72–77]. Further, isolated and non-isolated DC-DC power converters are the main configurations of DC-DC converters [26–29,78,79]. In isolated DC-DC power converter configurations, load and input are isolated electrically by utilizing transformer and coupled inductors [80–82].

In non-isolated DC-DC converters, the load and input sources share a common ground, but even with floating load input and load, still are not exactly electrically isolated [26], [71]. Both isolated and non-isolated DC-DC circuit are further classified into two configurations: 1. unidirectional and 2. bidirectional converters [71]. In isolated DC-DC power converter configurations, transformers are used to create isolation between input and load, but the transformer increases the size, cost, and losses [83–88]. Therefore, to minimize the transformer size, utilization of high

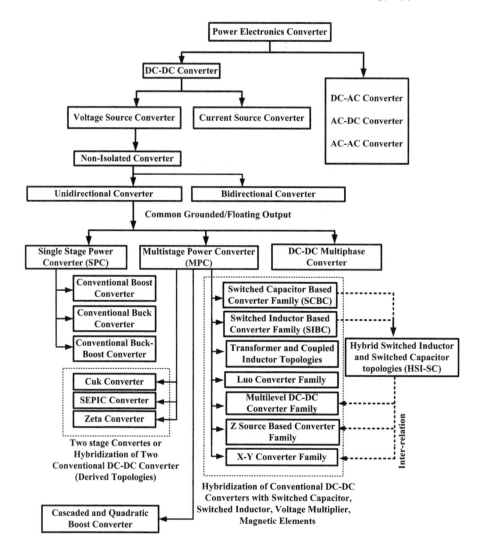

FIGURE 3.1 DC-DC Power Converter Configurations Classification.

frequency is the best option. However, the cost of the transformer is increased with frequency. Hence, non-isolated converters are more popular and provide the best option to overcome the drawback of the isolated converter. In bidirectional power converter configurations, power directed in both directions of the converter (input to the output port or output to an input port). However, in cases of unidirectional power converter configurations, power is directed in one direction (i.e., input to output port).

Both unidirectional and bidirectional non-isolated configurations are further classified into two new categories: common grounded and floating output [29,71]. Thus, non-isolated unidirectional standard ground converters (non-isolated UCGC) and unidirectional floating output converters (non-isolated UFOC) are two new configurations of non-isolated unidirectional power converters. Similarly, isolated unidirectional ground converters (isolated UGC) and unidirectional floating-output converters (isolated UFOC) are the two new configurations of non-isolated unidirectional power converters. After that, both UCGC and UFOC non-isolated power converters configurations classified into three main configurations: single-stage power converters, multistage power converters, and multiphase power converters. The block diagram of non-isolated UCGC, non-isolated UFOC, isolated UGC, and isolated UFOC single-stage DC-DC power converter configurations are shown in Figure 3.2(a)–(d), respectively [71]. The block diagram of non-isolated UCGC, non-isolated UFOC, isolated UGC, and isolated UFOC multistage DC-DC power converter configurations are shown in Figure 3.3(a)–(d) [71], respectively.

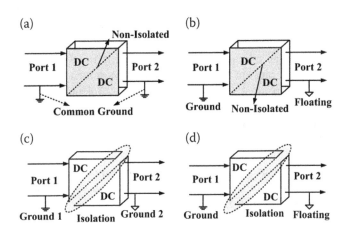

FIGURE 3.2 Block Diagram of Single-Stage DC-DC Power Converter Configurations: (a) Non-Isolated UCGC, (b) Non-Isolated UFOC, (c) Isolated UGC, (d) Isolated UFOC.

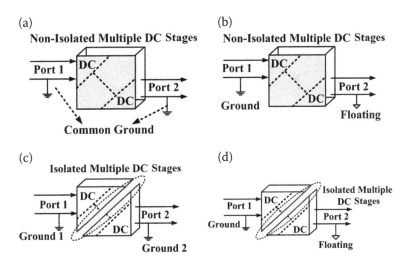

FIGURE 3.3 Block Diagram of Multistage DC-DC Power Converter Configurations: (a) Non-Isolated UCGC, (b) Non-Isolated UFOC, (c) Isolated UGC, (d) Isolated UFOC.

The functionality and voltage conversion ratio of DC-DC power converter configurations depend on the position, the number of reactive components, and the number of semiconductor uncontrolled/controlled switches in the power circuit.

3.3 CONFIGURATIONS OF SINGLE- AND TWO-STAGE DC-DC POWER CONVERTER

Buck, boost, and buck-boost converters are the single-stage conventional configurations of a DC-DC converter [71]. Buck converter configuration provides the output voltage lesser than the input voltage at all duty ratio [89,90]. Boost converter configuration provides an output voltage higher than the input voltage at all duty ratios. Buck-boost converter configurations provide a higher output voltage and lower input voltage at a duty ratio higher than 0.5 and lesser than 0.5, respectively. At a duty ratio of 0.5, buck-boost converter configurations provide output voltage equal to input voltage. Figure 3.4(a)–(c) depicts the power circuit of single-stage buck, boost, buck-boost converters with a common ground, respectively. Figure 3.5(a)–(c) depicts the power circuit of single-stage conventional buck, boost, buck-boost converters with floating output, respectively [26–29,71].

The buck-boost power converter configuration provides an inverting output voltage. Buck and boost power converter configurations provide a non-inverting output voltage. In recent times, many multistage power converter configurations are derived via hybridization of single-stage converters and using the front-end structure of the single-stage conventional converter [91–94]. Non-isolated unidirectional multistage power converter classifications and configurations are discussed in the next section of this chapter.

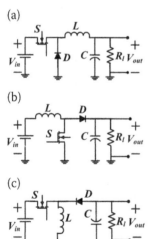

(a)

(b)

(c)

FIGURE 3.4   Power Circuit of Single-Stage DC-DC Power Converter Configurations: (a) Buck Converter, (b) Boost Converter, (c) Buck-Boost Converter.

(a)

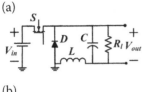

(b)

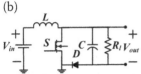

(c)

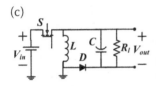

FIGURE 3.5 Power Circuit of Single-Stage DC-DC Power Converter Configurations with Floating Load: (a) Buck Converter, (b) Boost Converter, (c) Buck-Boost Converter.

3.4 CONFIGURATIONS OF MULTISTAGE DC-DC POWER CONVERTERS

Single-stage DC-DC power converter configurations are a good option for minimum-voltage step-up applications. However, the single-stage power converter configurations are not a feasible solution for high-voltage step-up applications due to voltage stress on switches and requirements of a higher duty ratio [95,96]. Recently, various DC-DC multistage power converter configurations are suggested to suit the voltage demand of loads and to make the power converter more efficient, reliable, small in size, and lightweight [26–29,71]. Multistage DC-DC power converter circuitry is derived by utilizing various boosting stages, along with the structure of single-stage DC-DC power converters [26,71]. The combinations of single-stage DC-DC power converters and boosting stages form a vast number of multistage DC-DC configurations. It is quite confusing and complicated to review and classify the multistage DC-DC power converters because each topology has its characteristics, features, and merits. Based on the review, in this section, the global scenario of newly multistaged DC-DC power converter configurations are presented to classify the multistage power converters. This section helps readers to understand the idea and various configurations of unidirectional multistage power converter configurations, and the many boosting stages with the advantages and disadvantages in terms of their reliability, cost, and applications. All the non-isolated DC-DC multistage converter configurations are classified into three categorizes based on the boosting capability and boosting stages. The multistage power converter configurations are categorized as follows:

- Low-voltage step-up multistage power converter configurations (two-stage or derived power converter configurations)
- Medium-voltage step-up multistage power converter configurations (cascaded or quadratic power converter configurations)

- High-voltage step-up multistage power converter configurations (based on switched inductor, switched capacitor, transformer, coupled inductor, Z source, quasi Z source, and voltage multiplier)

3.4.1 Low-Voltage Step-Up Multistage Power Converter Configurations

Two conventional single-stage power converter configurations are hybridized to design two-stage power converters or derived power converter configurations. Figure 3.6 depicts the classification of low-voltage step-up multistage power converter configurations [71]. Single-ended primary inductance converter (SEPIC), Cuk, and ZETA configurations are derived by the hybridization of two conventional single-stage DC-DC power converters. SEPIC, Cuk, and ZETA converter configurations are two-stage power converter configurations or derived configurations [91,92]. As a result, SEPIC, Cuk, and ZETA configurations are categorized into multistage power converter configurations. Many power electronics researchers claim that two-stage or derived multistage converters are conventional DC-DC converters. However, these configurations are derived by combining the features of two conventional single-stage DC-DC converters for advanced benefits and to overcome the drawback of the single-stage power converter. Various combinations of the conventional single-stage power converter are done to derive the low-voltage multistage power converter configuration as shown in Figure 3.7 [91,96].

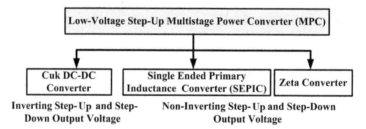

FIGURE 3.6 Classification of Low-Voltage Step-Up Multistage Power Converter Configurations.

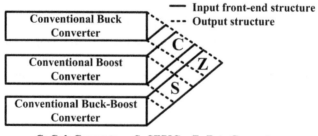

FIGURE 3.7 Derivation of Deriving the Low-Voltage Multistage Power Converter Configurations.

SEPIC configuration is derived by combining the features of boost and buck-boost power converters. Boost front ends and buck-boost load side structures are used to drive SEPIC and the non-inverting step-up/down voltage achieved at the output terminal. A Cuk converter configuration is derived by combining the features of boost and buck power converters. Boost front ends and buck load side structures are used to drive a Cuk converter and the inverting step-up/down voltage achieved at the output terminal.

A ZETA converter configuration is derived by combining the features of buck-boost and buck power converters. Buck-boost front ends and buck load side structures are used to drive a ZETA converter and the non-inverting step-up/down voltage achieved at the output terminal. A single-controlled switch, two capacitors, two inductors, and the single uncontrolled switch are required to design a power circuit of SEPIC, Cuk, and ZETA. Figure 3.8(a)–(c) depicts the power circuit of SEPIC, Cuk, and ZETA power converter configurations [71]. A conventional single-stage DC-DC, SEPIC, Cuk, and ZETA converter are not appropriate to achieve a high-voltage conversion ratio due to the requirement of a high duty ratio and highly rated reactive components and semiconductor devices.

3.4.2 MODERATE-VOLTAGE STEP-UP MULTISTAGE POWER CONVERTER CONFIGURATIONS

Single-stage and low-voltage step-up multistage power converters are not enough for the moderate- and high-voltage applications [91–97]. To get medium voltage, many single-stage converters are connected to design cascaded configuration of power

(a)

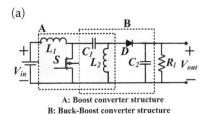

A: Boost converter structure
B: Buck-Boost converter structure

(b)

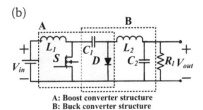

A: Boost converter structure
B: Buck converter structure

(c)

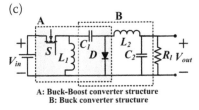

A: Buck-Boost converter structure
B: Buck converter structure

FIGURE 3.8 Power Circuit of Low-Voltage Step-Up Multistage Power Converter Configurations: (a) SEPIC, (b) Cuk, (c) ZETA.

converters. Figure 3.9 depicts generalized cascaded multistage power configurations [71,98]. Using cascaded multistage power converter configurations and a moderate-voltage conversion ratio is achieved but compromised in the robustness because of the requirement of several controlled and uncontrolled switches, reactive components, and complex circuitry to control power switches [96–100]. A high duty ratio is used to control the switch of the first stage and the low duty ratio is used to control switches of the other stages to minimize the switching losses of the converter. A cascaded approach is achieved using Cuk converter configuration but requires numerous components, which increases the losses. Hence, efficiency is decreased [101,102].

Quadratic boost converter (QBC) configurations are derived from overcoming several switch drawbacks of cascaded converter configurations [101–104]. Figure 3.10 depicts the generalized quadratic multistage power configuration [71,98]. Further, by using a more significant number of diodes and controlled switches, the QBC approach employed N-stages to increase a voltage conversion ratio. The total voltage conversion ratio is a product of a voltage conversion ratio of the individual stage, but the drawback is still available in QBC configurations. Switch stress, low efficiency, and complexity are the major drawbacks of QBC configurations. It requires highly rated devices and components because of high voltage stress across the devices. Thus, the cost of the QBC configuration increased.

In literature, several approaches are elaborated to increase the efficiency with low cost. In [106], two controlled semiconductor switches are used to develop a three-level step-up Quadratic Boost Converter (QBC) circuit for high-voltage applications. However, the

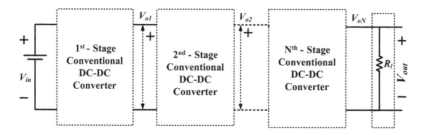

FIGURE 3.9 Generalized Cascaded Multistage Power Configuration.

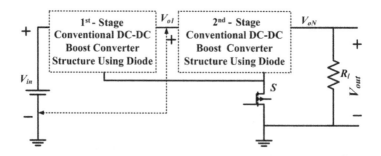

FIGURE 3.10 Generalized Quadratic Multistage Power Configuration.

drawback lying with the reactive components still exists, and thus these step-up circuits are limits to medium-voltage applications.

3.4.3 HIGH-VOLTAGE STEP-UP MULTISTAGE POWER CONVERTER CONFIGURATIONS

The front end, load and structure, or full single-stage power converters, along with boosting stages like a switched inductor, switched capacitor, Z source, quasi Z source, and Cockcroft voltage multiplier are used to design high-voltage step-up multistage power converter configurations. Various switched capacitor cells are addressed by using diodes and capacitors.

Figure 3.11(a)–(o) depicts the recently addressed switched capacitor cells, which are used to design switched-capacitor-based multistage power converter configurations [107–131]. Various switched inductor cells are addressed by using diodes and inductors. To get a higher voltage conversion ratio, a hybrid combination of a switched inductor and switched capacitor (HSI-SC) is also addressed. For simplicity, in this chapter, HSI-SC cells are considered a category of switched inductor cells. Figure 3.12(a)–(t) depicts the recently addressed switched inductor and HSI-SC cells that are used to design a switched inductor and HSI-SC-based multistage power converter configuration [71,132–151].

High-voltage step-up multistage power converter configurations are classified in the following five categories:

- Multistage switched-capacitor-based power converter family
- Multistage switched inductor power converter family or multistage HSI-SC-based power converter family
- A coupled inductor or transformer-based converter family
- Luo DC-DC converter family
- Z source–based converter family
- Cockcroft Walton voltage multiplier-based multilevel DC-DC converter family

3.5 MULTISTAGE SCBPC FAMILY (MULTISTAGE SWITCHED-CAPACITOR-BASED POWER CONVERTER FAMILY)

In recent times, switched capacitor (SC) cells are addressed to attain a high-voltage conversion ratio for several step-up applications [107–131]. A multistage SCBPC power converter family (SC-based multistage power converter configurations) is well-liked because of its modularity, uncomplicated structure, and possibilities for integration in a monolithic structure. To achieve a high-voltage conversion ratio, it performs all the capacitors of SC parallel charging and series discharging operation. The voltage conversion ratio of multistage SCBPC configurations depends on the arrangement and number of switched-capacitor stages used in the converter. A few configurations of multistage SCBPC configurations are called charge pump networks because one capacitor's energy is transferred to

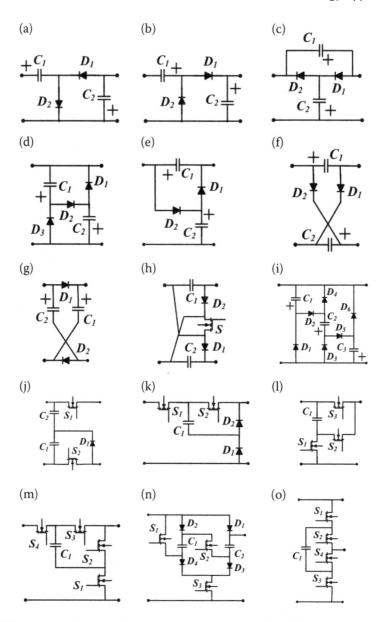

FIGURE 3.11   (a)–(o) Recently Addressed Switched Capacitor Cells.

another. SC cells provide a good agreement for low-weighted step-up power converters in the absence of an inductor and transformer. Many multistage SCBPC configurations are possible by combining the features of SC cells and traditional DC-DC or derived DC-DC converters (SEPIC, Cuk, and ZETA) [107–131].

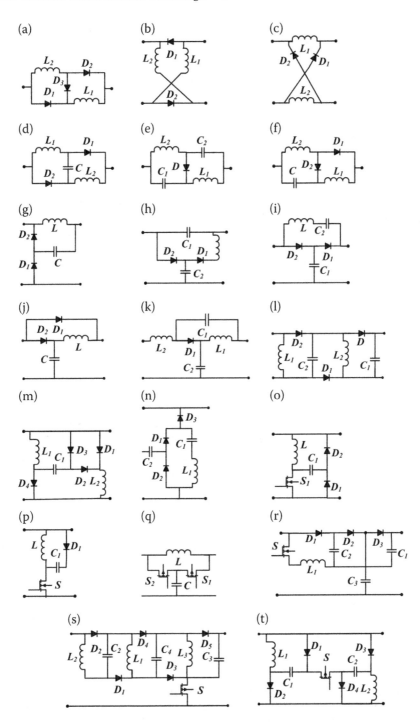

FIGURE 3.12 Recently Addressed (a)–(c) Switched Inductor Cells and (d)–(t) HSI-SC Cells.

3.5.1 MULTISTAGE SCBPC WITH FRONT-END STRUCTURE OF BOOST AND BUCK-BOOST CONVERTERS

In the last decades, several multistage SCBPC configurations with front-end structures of boost and buck-boost converters are addressed in literature. In this section, popular multistage SCBPC configurations are reviewed and discussed in detail. The power circuit of a three-switch boost converter (TSBC) configuration is shown in Figure 3.13(a), which is derived by combining the front-end structure of boost converter (FES-BC) and SC cell [113]. Single switches, two capacitors, two diodes, and a single inductor are required in the designed power circuit of a TSBC configuration. They are inverting output voltage received at the output terminal of the converter. The TSBC configuration is operated in a continuous conduction mode (CCM) and a discontinuous conduction mode (DCM). The voltage conversion ratio of a TSBC configuration for CCM and DCM is provided in equations (3.1) and (3.2), respectively. The converter operates in the CCM when $k > \Delta(1 - \Delta)^2$ and operates in the DCM when $k < \Delta(1 - \Delta)^2$.

$$G_{TSBC} = \frac{V_{out}}{V_{in}} = \frac{-1}{1 - \Delta} = \frac{-1}{1 - \frac{T_{on}}{T}} = \frac{1}{f_s T_{on} - 1} \qquad (3.1)$$

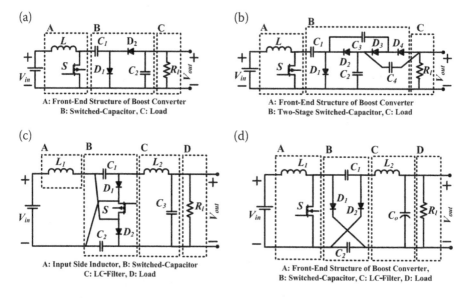

(a)

A: Front-End Structure of Boost Converter
B: Switched-Capacitor, C: Load

(b)

A: Front-End Structure of Boost Converter
B: Two-Stage Switched-Capacitor, C: Load

(c)

A: Input Side Inductor, B: Switched-Capacitor
C: LC-Filter, D: Load

(d)

A: Front-End Structure of Boost Converter,
B: Switched-Capacitor, C: LC-Filter, D: Load

FIGURE 3.13 Power Circuit: (a) Three-Switch Boost Converter (TSBC) Configuration, (b) Extended Structure of TSBC Configuration, (c) Inverting Output Switched-Capacitor Converter (IO-SCC) Configuration, (d) Non-Inverting Output Switched-Capacitor Converter (NO-SCC) Configuration, (e) Non-Inverting Switched-Capacitor Boost Converter (N-SCBC) Configuration, (f) Inverting Switched-Capacitor Boost Converter (I-SCBC) Configuration.

$$G_{TSBC} = \frac{V_{out}}{V_{in}} = -\frac{1 + \sqrt{1 + \frac{4\Delta^2}{k}}}{2}, \; k = \frac{2L}{T_s R_l} \tag{3.2}$$

When the load resistance is RL, ON time of the switch is T_{on}, the switching frequency is $f_s = 1/T_s$. The TSBC configuration is easily extendable to achieve a high-voltage converter ratio by adding additional SC cells. Figure 3.13(b) depicts the extended structure of the TSBC configuration. To add one stage of SC into a TSBC configuration, two capacitors and two diodes are required. In (3.3), the voltage conversion ratio of extended TSBC configuration is provided.

$$G_{ext-TSBC} = \frac{V_{out}}{V_{in}} = \frac{-2}{1 - \Delta} = \frac{-2}{1 - \frac{T_{on}}{T}} = \frac{2}{f_s T_{on} - 1} \tag{3.3}$$

The power circuit of the inverting output switched-capacitor converter (IO-SCC) configuration with an input inductor is shown in Figure 3.13(c) [114]. Switch-capacitor stage, input inductor stage, and LC filter are the three stages required to design IO-SCC configuration. Three capacitors, two inductors, two diodes, and a single-controlled switch are required to design an IO-SCC configuration. An IO-SCC configuration provides a negative output, and the voltage conversion ratio is provided in (3.4).

$$G_{IOSC} = \frac{V_{out}}{V_{in}} = \frac{-(1 + \Delta)}{1 - \Delta} = \frac{-\left(1 + \frac{T_{on}}{T}\right)}{1 - \frac{T_{on}}{T}} = \frac{f_s T_{on} + 1}{f_s T_{on} - 1} \tag{3.4}$$

The power circuit of the non-inverting output switched-capacitor converter (NO-SCC) configuration with an input inductor is shown in Figure 3.13(d) [114]. Switched-capacitor, input inductor, and LC filter are the three stages required to design an IO-SCC configuration. Three capacitors, two inductors, two diodes, and single controlled switch are required to design a NO-SCC configuration. A NO-SCC configuration provides a positive output, and the voltage conversion ratio is provided in (3.5).

$$G_{NOSC} = \frac{V_{out}}{V_{in}} = \frac{(1 + \Delta)}{1 - \Delta} = \frac{1 + \frac{T_{on}}{T}}{1 - \frac{T_{on}}{T}} = \frac{f_s T_{on} + 1}{1 - f_s T_{on}} \tag{3.5}$$

The power circuit of the non-inverting switched-capacitor boost converter (N-SCBC) configuration with minimal voltage stress of the switch is shown in Figure 3.13(e) [115,116]. FES-BC, switched capacitor, and LC filter stages are required to design a N-SCBC configuration. Three capacitors, two inductors, two diodes, and a single controlled switch are required to design N-SCBC configurations. A N-SCBC configuration provides a positive output, and the voltage conversion ratio is provided in (3.6).

$$G_{N-SCBC} = \frac{V_{out}}{V_{in}} = \frac{1 + \Delta}{1 - \Delta} = \frac{1 + \frac{T_{on}}{T}}{1 - \frac{T_{on}}{T}} = \frac{T_{on}f_s + 1}{1 - T_{on}f_s} \qquad (3.6)$$

The power circuit of the inverting switched-capacitor boost converter (I-SCBC) configuration with minimal voltage stress of switch is shown in Figure 3.13(f) [115,116]. FES-BC, switch capacitor stage, and LC filter stages are required to design a N-SCBC configuration. Three capacitors, two inductors, two diodes, and a single controlled switch are required to design N-SCBC configurations. A N-SCBC configuration provides a negative output, and the voltage conversion ratio is provided in (3.7).

$$G_{I-SCBC} = \frac{V_{out}}{V_{in}} = \frac{-(1 + \Delta)}{1 - \Delta} = \frac{-\left(1 + \frac{T_{on}}{T}\right)}{1 - \frac{T_{on}}{T}} = \frac{-(1 + f_s T_{on})}{1 - f_s T_{on}} \qquad (3.7)$$

The power circuit of two-stage switched-capacitor boost converter (TSC-BC) configuration is shown in Figure 3.14(a) [113]. This configuration has low voltage stress across the switch, and is also called as high voltage three-switch Cuk converter (HVTS-CC) configuration because input and output side characteristics are similar to a Cuk converter. FES-BC, SC, and LC filter stages are required to design TSC-BC or HVTS-CC configurations. This converter provides inverting output voltage; the voltage conversion ratio is given in (3.8).

$$G_{TSC-BC} = \frac{V_{out}}{V_{in}} = \frac{-(1 + \Delta)}{1 - \Delta} = \frac{-\left(1 + \frac{T_{on}}{T}\right)}{1 - \frac{T_{on}}{T}} = \frac{1 + T_{on}f_s}{f_s T_{on} - 1} \qquad (3.8)$$

The power circuit of a derived high voltage Cuk converter (DHV-CC) configuration is shown in Figure 3.14(b) [119]. This converter provides inverting output and with only an input inductor. FES-BC, SC, and C-filter stages are used to design the DHV-CC configuration. FES-BC and SC combine to form a Cuk input side structure. Three capacitors, three diodes, one inductor, and one control switch are required to design DHV-CC configurations. A double voltage conversion ratio of low voltage stress on the switch is achieved, compared to the conventional boost converter. The voltage conversion ratio of DHV-CC is provided in (3.9).

$$G_{DHV-CC} = \frac{V_{out}}{V_{in}} = \frac{-2}{1 - \Delta} = \frac{-2}{1 - \frac{T_{on}}{T}} = \frac{2}{T_{on}f_s - 1} \qquad (3.9)$$

The power circuit of four stages of a derived high voltage SEPIC (DHV-SEPIC) configuration is shown in Figure 3.14(c). It combines SC with the power circuit of SEPIC to derive a DHV-SEPIC configuration [117–119]. Intermediate, FES-BC, SC, and C-filter stages are required to design a DHV-SEPIC configuration. Four

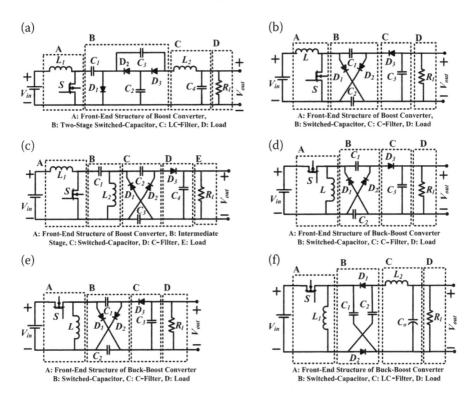

FIGURE 3.14 Power Circuits: (a) Two-Stage Switched-Capacitor Boost Converter (TSC-BC) Configurations, (b) Derived High Voltage Cuk Converter (DHV-CC) Configuration, (c) Derived High Voltage SEPIC (DHV-SEPIC) Configuration, (d) Derived High Voltage ZETA (DHV-ZETA) Converter Configuration, (e) Derived High Voltage Inverting ZETA (DHVI-ZETA) Converter Configuration, (f) Diode Assist Switched-Capacitor Converter (DA-SCC) Configuration.

capacitors, two inductors, two diodes, and a single switch are required to design stages of the DHV-SEPIC configuration. This configuration provides a high non-inverting output voltage with low voltage stress across the switch, compared to the conventional boost converter. The voltage conversion ratio of DHV-SEPIC is provided in (3.10). We can change the capacitor polarity and diode direction to get an inverting output voltage.

$$G_{DHV-SEPIC} = \frac{V_{out}}{V_{in}} = \frac{2-\Delta}{1-\Delta} = \frac{2 - \frac{T_{on}}{T}}{1 - \frac{T_{on}}{T}} = \frac{T_{on}f_s - 2}{T_{on}f_s - 1} \qquad (3.10)$$

The power circuit of three stages of the derived high voltage ZETA (DHV-ZETA) converter configuration is shown in Figure 3.14(d) [117–119]. It combines SC with the power circuit of ZETA to derive a DHV-ZETA configuration and an output side inductor of ZETA charged by the diode. FES-BBC, SC, and C-filter stages are required to design a DHV-ZETA configuration. A ZETA structure is formed by combining FES-BBC and SC. Three capacitors, one inductor, three diodes, and a single switch are required to design stages of the DHV-ZETA configuration. This configuration provides a high non-inverting output voltage with low voltage stress across switch compared to the conventional boost converter. A floating control switch is used at the input side of the inductor. The voltage conversion ratio of a DHV-ZETA for CCM and DCM is provided in equations (3.11) and (3.12), respectively.

$$G_{DHV-ZETA} = \frac{V_{out}}{V_{in}} = \frac{1 + \Delta}{1 - \Delta} = \frac{1 + \frac{T_{on}}{T}}{1 - \frac{T_{on}}{T}} = \frac{T_{on}f_s + 1}{1 - T_{on}f_s} \qquad (3.11)$$

$$G_{DHV-ZETA} = \frac{V_{out}}{V_{in}} = -\frac{1 + \sqrt{1 + \frac{4\Delta^2}{k}}}{2}, \ k = \frac{2L}{T_s R_l} \qquad (3.12)$$

The power circuit of three stages of a derived high voltage inverting ZETA (DHVI-ZETA) converter configuration shown in Figure 3.14(e) [117–119]. It combines SC with the power circuit of ZETA to derive a DHVI-ZETA configuration and output side inductor of ZETA charged by the diode. FES-BBC, SC, and C-filter stages are required to design a DHVI-ZETA configuration. A ZETA structure is formed by combining FES-BBC and SC. Three capacitors, one inductor, three diodes, and a single switch are required to design stages of the DHVI-ZETA configuration. This configuration provides an inverting high output voltage with low voltage stress across a switch, compared to the conventional boost converter. There are floating control switches used at the input side of the inductor. The voltage conversion ratio of DHVI-ZETA for CCM and DCM is provided in equations (3.13) and (3.14), respectively.

$$G_{DHVI-ZETA} = \frac{V_{out}}{V_{in}} = -\frac{2 - \Delta}{1 - \Delta} = -\frac{2 - \frac{T_{on}}{T}}{1 - \frac{T_{on}}{T}} = \frac{2 - T_{on}f_s}{T_{on}f_s - 1} \qquad (3.13)$$

$$G_{DHVI-ZETA} = \frac{V_{out}}{V_{in}} = -\left(1 + \sqrt{1 + \frac{\Delta^2}{k}}\right), \ k = \frac{2L}{T_s R_l} \qquad (3.14)$$

The power circuit of a diode assist switched-capacitor converter (DA-SCC) configuration is shown in Figure 3.14(f). The DA-SCC configuration provides inverting output voltage, and it is derived by combining the features of a SC and ZETA converter [117–119]. FES-BBC, SC, and LC-filter stages are required to design a DA-SCC configuration. Three capacitors, two inductors, two diodes, and a single switch are required to design stages of the DA-SCC configuration.

3.5.2 Multistage SCBPC Configurations without FES-BC and FES-BBC

Many multistage SCBPC configurations without FES-BC and FES-BBC are suggested in literature [71,120–131]. These power converter configurations are suggested for high step-up and high step-down applications and configurations, and also provide a solution for low-weight photovoltaic applications. The power circuits of intermediate boost switched-capacitor converter (IB-SCC) configurations are shown in Figure 3.15(a) [120]. This configuration provides a high voltage at the output side. SC is an input stage and a boost converter is used in the middle (intermediate stage) of the converter. Three capacitors, single inductor, five diodes, and four control switches are required to design stages of the IB-SCC configuration. The non-inverting output voltage received and the voltage conversion ratio of IB-SCC configuration are provided in (3.15).

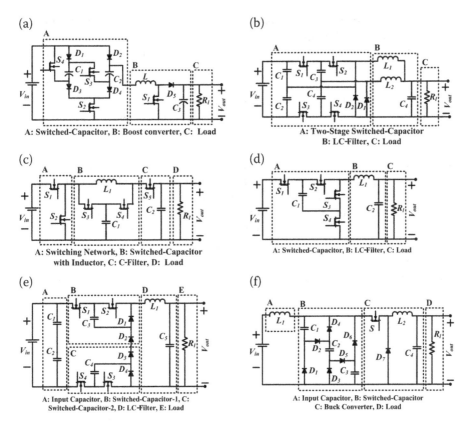

FIGURE 3.15 Power Circuits: (a) Intermediate Boost Switched-Capacitor Converter (IB-SCC) Configuration, (b)–(f) Multistage Switched-Capacitor-Based Step-Down Power Converter Configurations.

$$G_{IB-SCC} = \frac{V_{out}}{V_{in}} = \frac{3 - 2\Delta}{1 - \Delta} = \frac{3 - 2\frac{T_{on}}{T}}{1 - \frac{T_{on}}{T}} = \frac{2T_{on}f_s - 3}{T_{on}f_s - 1} \qquad (3.15)$$

The power circuit of the multistage switched-capacitor-based step-down power converter configurations are shown in Figure 3.15(b)–(f). It suits these converter configurations well for low-voltage applications [121–131].

3.5.3 MULTISTAGE SCBPC CONFIGURATIONS WITH FRONT-END STRUCTURE OF QUADRATIC BOOST CONVERTER (FES-QBC)

A front-end structure of quadratic boost converter (FES-QBC) is used with SC to attain a high-voltage conversion ratio. Based on the intermediate stage and filter, four new multistage power converter configurations are as follows:

- Multistage switched-capacitor quadratic boost converter (MSC-QBC) configuration with C-filter
- Multistage switched-capacitor quadratic boost converter (MSC-QBC) configuration with LC-filter
- Multistage switched-capacitor quadratic boost converter (MSC-QBC) configuration with intermediate stage and C-filter
- Multistage switched-capacitor quadratic boost converter (MSC-QBC) configuration with intermediate stage and LC-filter

The power circuit of a MSC-QBC configuration with C-filter is shown in Figure 3.16(a) [71] and the voltage conversion ratio is provided in (3.16). The power circuit of a MSC-QBC configuration with LC-filter is shown in Figure 3.16(b) [71] and the voltage conversion ratio is provided in (3.17). The power circuit of a MSC-QBC configuration with intermediate stage and C-filter is shown in Figure 3.16(c) [71] and the voltage conversion ratio is provided in (3.18). The power circuit of a MSC-QBC configuration with an intermediate stage and LC-filter is shown in Figure 3.16(d) [71] and the voltage conversion ratio is provided in (3.19).

$$G_{\text{MSC-QBC,C-filter}} = \frac{V_{out}}{V_{in}} = \frac{2}{(1 - \Delta)^2} = \frac{2}{(1 - T_{on}f_s)^2} \qquad (3.16)$$

$$G_{\text{MSC-QBC,LC-filter}} = \frac{V_{out}}{V_{in}} = \frac{1 + \Delta}{(1 - \Delta)^2} = \frac{1 + T_{on}f_s}{(1 - T_{on}f_s)^2} \qquad (3.17)$$

$$G_{\text{MSC-QBC, stage and C-filter}} = \frac{V_{out}}{V_{in}} = \frac{2 - \Delta}{(1 - \Delta)^2} = \frac{2 - T_{on}f_s}{(1 - T_{on}f_s)^2} \qquad (3.18)$$

(a)

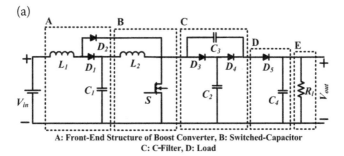

A: Front-End Structure of Boost Converter, B: Switched-Capacitor
C: C-Filter, D: Load

(b)

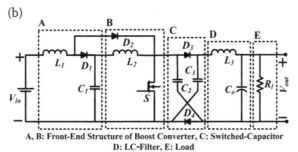

A, B: Front-End Structure of Boost Converter, C: Switched-Capacitor
D: LC-Filter, E: Load

(c)

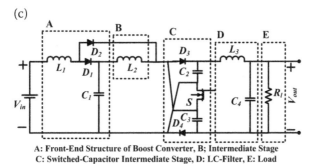

A: Front-End Structure of Boost Converter, B; Intermediate Stage
C: Switched-Capacitor Intermediate Stage, D: LC-Filter, E: Load

(d)

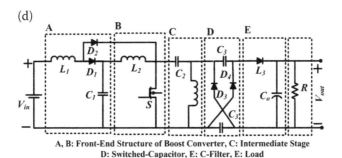

A, B: Front-End Structure of Boost Converter, C: Intermediate Stage
D: Switched-Capacitor, E: C-Filter, E: Load

FIGURE 3.16 Power Circuits: (a) MSC-QBC Configuration with C-Filter, (b) MSC-QBC Configuration with LC-Filter, (c) MSC-QBC Configuration with Intermediate Stage and C-Filter, (d) MSC-QBC Configuration with Intermediate Stage and LC-Filter.

$$G_{\text{MSC}-\text{QBC, stage and LC}-\text{filter}} = \frac{V_{out}}{V_{in}} = \frac{1 + \Delta}{(1 - \Delta)^2} = \frac{1 + T_{on}f_s}{(1 - T_{on}f_s)^2} \qquad (3.19)$$

Based on the front-end structure, the classification of the multistage switched-capacitor-based power converter (M-SCBPC) family is given in Figure 3.17.

3.6 MULTISTAGE SIBPC FAMILY (MULTISTAGE SWITCHED-INDUCTOR-BASED POWER CONVERTER FAMILY)

To attain a high-voltage conversion ratio, multistage switched-inductor (SI) based power converter configurations are a recent trend in power electronics converters and require a smaller number of components [132,133]. In SI, two or more inductors are performing parallel charging and series discharging operations to attain a high conversion ratio. The configuration is simple and to reduce power converter size and weight, two or more inductors with the same rating can be integrated on a single core. To achieve more benefits, the hybridization of SI and SC (HSI-SC cell) is another well-accepted solution [134–151]. Recently, various M-SIBPC configurations are addressed in literature for step-up or step-down applications [134–151]. In this section, SI- and HIS-SC-based power converter configurations are explained. The power circuit of the basic configuration of a switched-inductor boost converter (Basic-SIBC) is shown in Figure 3.18(a) [132,134].

The feature of the conventional boost converter and SI combined to derive a Basic-SIBC configuration. Two identical inductors with the same rating, four diodes, one capacitor, and one control switch are required to design a Basic-SIBC configuration. Switching stage, SI, and C-filter stages are required to design a Basic-SIBC configuration. The voltage conversion ratio of a Basic-SIBC is provided in (3.20).

$$G_{Basic-SIBC} = \frac{V_{out}}{V_{in}} = \frac{1 + \Delta}{1 - \Delta} = \frac{1 + \frac{T_{on}}{T}}{1 - \frac{T_{on}}{T}} = \frac{1 + T_{on}f_s}{1 - T_{on}f_s} \qquad (3.20)$$

The power circuit of a hybrid quadratic boost converter (Hybrid-QBC) configuration is based on a HSI-SC cell, as shown in Figure 3.18(b) [135]. A high-voltage

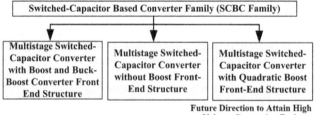

FIGURE 3.17 Classification of Multistage Switched-Capacitor-Based Power Converter (M-SCBPC) Family.

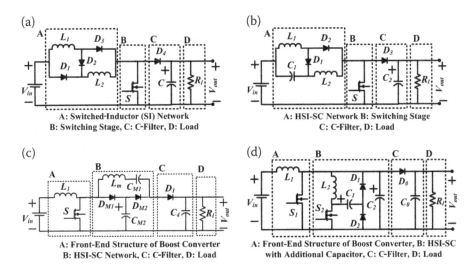

FIGURE 3.18 Power Circuits: (a) Basic-SIBC Configuration, (b) Hybrid Quadratic Boost Converter (Hybrid-QBC) Configuration, (c) HSI-SC-Based Boost Converter, (d) Soft Switching High Step-Up Converter Configuration.

conversion ratio is achieved with a low buffer capacitor stress, and configuration is designed by combining the feature of a HSI-SC with a conventional boost converter. Switching stage, HSI-SC, and C-filter stages are required to design the converter. For low voltage across the capacitor, it suits this configuration for a high-voltage conversion ratio compared to conventional boost and quadratic boost converters. The voltage conversion ratio of a Hybrid-QBC configuration is provided in (3.21).

$$G_{H-QBC} = \frac{V_{out}}{V_{in}} = \frac{1}{(1 - \Delta)^2} = \frac{1}{\left(1 - \frac{T_{on}}{T}\right)^2} = \frac{1}{(1 - T_{on}f_s)^2} \qquad (3.21)$$

The power circuit of a HSI-SC-based boost converter is shown in Figure 3.18(c) [136]. In this circuit, the HIS-SC cell acts as a multiplier and is adjusted in the center of the boost converter. Zero current switching (ZCS) is possible to turn ON the switch and the effect of reverse recovery for the diode is also reduced. The configuration is also possible to operate at high frequency and commutation losses are also reduced. The converter configuration is also extendable to N-stages by the addition of a more significant number of stages of the multiplier. To achieve high voltage at the output with reducing ripples, the HSI-SC intermediate stage is used, along with a C-filter. (3.22), (3.23). The voltage conversion ratio of 1-stage and N-stage HSI-SC-based boost converter is provided in equations (3.22) and (3.23), respectively.

$$G_{\text{one-stage}} = \frac{V_{out}}{V_{in}} = \frac{2}{1 - \Delta} = \frac{2}{1 - \frac{T_{on}}{T}} = \frac{2}{1 - T_{on}f_s} \qquad (3.22)$$

$$G_{N-\text{stage}} = \frac{V_{out}}{V_{in}} = \frac{N+1}{1-\Delta} = \frac{N+1}{1-\frac{T_{on}}{T}} = \frac{N+1}{1-T_{on}f_s} \qquad (3.23)$$

The power circuit of a soft switching high step-up converter configuration is shown in Figure 3.18(d) [137]. The HIS-SC stage is used with the additional capacitor to achieve a high-voltage conversion ratio and high efficiency. FES-BC, capacitor-based HSI-SC, and C-filter stages are used to design a converter configuration to obtain high-voltage at the output. However, complex circuitry to design the driver is required. But, it does not require additional ground separation for both switches. The voltage conversion ratio of a soft switching high step-up converter configuration is provided in equation (3.24).

$$G_{\text{soft switch converter}} = \frac{V_{out}}{V_{in}} = \frac{1}{1-(\Delta_1+\Delta_2)} = \frac{1}{1-f_s(T_{on1}+T_{on2})} \qquad (3.24)$$

The power circuit of a step-up ultra-converter (SUC) configuration is shown in Figure 3.19(a) [138]. SUC configurations are designed by combining the features of HSI-SC and SC with a buck-boost converter to achieve a high-voltage conversion

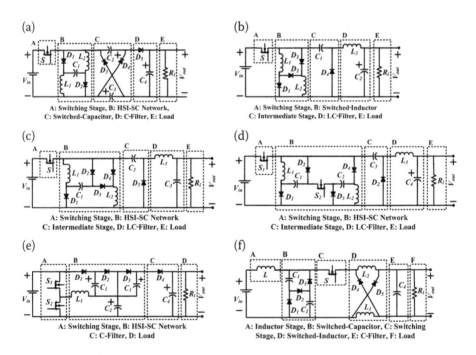

FIGURE 3.19 Power Circuits: (a) Step-Up Ultra-Converter (SUC) Configuration, (b) SI-Based Hybrid Positive Output Converter Configuration, (c) HSI-SC-Based Positive Output Self-Lift Converter Configuration, (d) HSI-SC-Based Hybrid Positive Output Double Self-Lift Converter Configuration, (e) Triple Mode Switched-Capacitor Converter (TM-SCC) Configuration, (f) HSI-SC-Based Buck Converter Configuration.

ratio. Semiconductor devices with low ratings can design SUC configurations because of the low voltage across the switch and diode. Hence, the Schottky diode is used to design the SUC configuration. The voltage conversion ratio of the SUC configuration is provided in (3.25).

$$G_{SUC} = \frac{V_{out}}{V_{in}} = \frac{3 + \Delta}{1 - \Delta} = \frac{3 + \frac{T_{on}}{T}}{1 - \frac{T_{on}}{T}} = \frac{3 + T_{on}f_s}{1 - T_{on}f_s} \tag{3.25}$$

The power circuit of a SI-based hybrid positive output converter configuration is shown in Figure 3.19(b) [139]. The voltage conversion ratio of a SI-based hybrid positive output converter is provided in (3.26). The power circuit of a HSI-SC-based positive output self-lift converter configuration is shown in Figure 3.19(c). The voltage conversion ratio of a HSI-SC-based positive output self-lift converter is provided in (3.27). The power circuit of a HSI-SC-based hybrid positive output double self-lift converter configuration is shown in Figure 3.19(d). The voltage conversion ratio of a hybrid positive output double self-lift converter configuration is provided in (3.28). HIS-SC, switching stage, intermediate, and LC-filter stages are used to design these configurations.

$$G_{SI} = \frac{V_{out}}{V_{in}} = \frac{\Delta + \Delta^2}{1 - \Delta} = \frac{T_{on}f_s + T_{on}^2f_s^2}{1 - T_{on}f_s} \tag{3.26}$$

$$G_{Self-Lift} = \frac{V_{out}}{V_{in}} = \frac{2\Delta}{1 - \Delta} = \frac{2T_{on}f_s}{1 - T_{on}f_s} \tag{3.27}$$

$$G_{Double-Self-Lift} = \frac{V_{out}}{V_{in}} = \frac{3\Delta - \Delta^2}{1 - \Delta} = \frac{3T_{on}f_s - T_{on}^2f_s^2}{1 - T_{on}f_s} \tag{3.28}$$

The power circuit of the triple mode switched-capacitor converter (TM-SCC) configuration is shown in Figure 3.19(e) [140]. The resonant tank used for ZCS operation and the current spike is minimized for ZCS. The TM-SCC configuration is divided into three stages: HSI-SC stage, switching stage, and filter stage. The power circuit of a HSI-SC-based buck converter configuration is shown in Figure 3.19(f) [131,140,141].

$$G_{Buck} = \frac{V_{out}}{V_{in}} = \frac{\Delta}{(2 - \Delta)^2} = \frac{T_{on}}{\left(2 - \frac{T_{on}}{T}\right)^2} = \frac{T_{on}f_s}{(2 - T_{on}f_s)^2} \tag{3.29}$$

To design the power circuit of the buck converter, three capacitors, three inductors, five diodes, and one control switch are required. This converter configuration is divided into four stages: switching stage, inductor stage at the input side, switched-capacitor stage, and switched-inductor stage. This converter provides a large step-down voltage at the output of the converter with reasonable efficiency and the conversion ratio is provided in (3.29).

The power circuit of a HSI-SC-based quadratic buck converter configuration is shown in Figure 3.20(a) [121]. The HIS-SC cell is used at the input side of a buck converter to achieve a large step-down voltage conversion ratio. The voltage conversion ratio of a HSI-SC-based quadratic buck converter configuration is provided in (3.30).

The power circuit of a multiplier and HSI-SC cell-based buck converter is shown in Figure 3.20(b). The reverse recovery problem of current is reduced by a reactive component arrangement in the converter configuration [141–144]. An electro-magnetic interference (EMI) problem is also overcome because the HSI-SC circuitry acts as a regenerative clamping circuit. The power circuit of the raw SI buck converter is shown in Figure 3.20(c). This configuration is designed by utilizing a SI cell in a conventional buck converter. The voltage conversion ratio of the raw SI buck converter is provided in (3.31) [141–144].

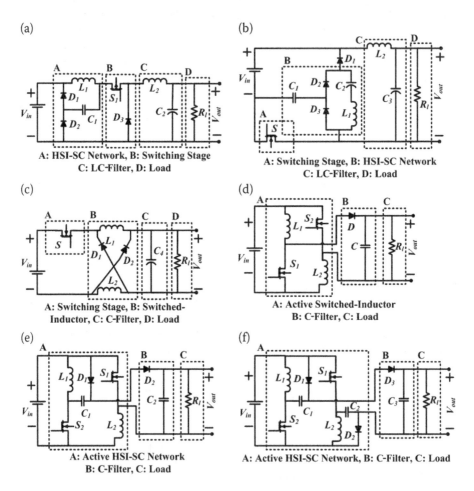

FIGURE 3.20 Power Circuits: (a) HSI-SC-Based Quadratic Buck Converter, (b) Multiplier and HSI-SC Cell-Based Buck Converter, (c) Basic SI Buck Converter, (d) Converter-A, (e) Converter-B, (f) Converter-C.

$$G_{HSI-SC-Buck-converter} = \frac{V_{out}}{V_{in}} = \Delta^2 = T_{on}^2 f_s^2 \qquad (3.30)$$

$$G_{Basic-SI-Buck-Converter} = \frac{V_{out}}{V_{in}} = \frac{\Delta}{2 - \Delta} = \frac{T_{on}f_s}{2 - T_{on}f_s} \qquad (3.31)$$

Figure 3.20(d)–(f) depicts the power circuit of three new DC-DC converter configurations (Converter-A, Converter-B, and Converter-C) without a transformer; two switches are used in the switched inductor instead of three diodes [134]. The voltage lifting technique is used to attain a higher voltage conversion ratio. Equations (3.32)–(3.34) provide the voltage conversion ratio of Converter-A, Converter-B, and Converter-C, respectively.

$$G_{Converter-A} = \frac{V_{out}}{V_{in}} = \frac{1 + \Delta}{1 - \Delta} = \frac{1 + \frac{T_{on}}{T}}{1 - \frac{T_{on}}{T}} = \frac{1 + T_{on}f_s}{1 - T_{on}f_s} \qquad (3.32)$$

$$G_{Converter-B} = \frac{V_{out}}{V_{in}} = \frac{2}{1 - \Delta} = \frac{2}{1 - \frac{T_{on}}{T}} = \frac{2}{1 - T_{on}f_s} \qquad (3.33)$$

$$G_{Converter-C} = \frac{V_{out}}{V_{in}} = \frac{3 + \Delta}{1 - \Delta} = \frac{3 + \frac{T_{on}}{T}}{1 - \frac{T_{on}}{T}} = \frac{3 + T_{on}f_s}{1 - T_{on}f_s} \qquad (3.34)$$

3.7 COUPLED INDUCTOR OR TRANSFORMER-BASED CONVERTER FAMILY

Coupled inductors and transformers are another solution for non-isolated and isolated converter configurations to achieve a high-voltage conversion ratio. Voltage lift technique, magnetic coupling, and built-in transformers are used in these configurations to achieve a high-voltage conversion ratio [26–29]. To attain a high-voltage conversion ratio, the energy stored in one winding of a built-in transformer is transferred to another winding through magnetic coupling. The coupled inductor is another viable solution to avoid a transformer in a DC-DC converter configuration for low cost, low weight, and reduced size. Leakage inductance is used to limit the diode current falling rate and to reduce the problem of recovery. Recently, untapped and tapped inductive couplings are used in the DC-DC converter to achieve a high-voltage conversion ratio [152–154].

The power circuit of a standard ground-coupled inductor is based on a step-up flying converter configuration shown in Figure 3.21(a). The voltage conversion ratio is in (3.35) [135]. This converter provides a high-voltage conversion ratio, but voltage stress across the switch is higher and efficiency is reduced for leakage inductance. The voltage conversion ratio of this converter configuration depends on the coupling

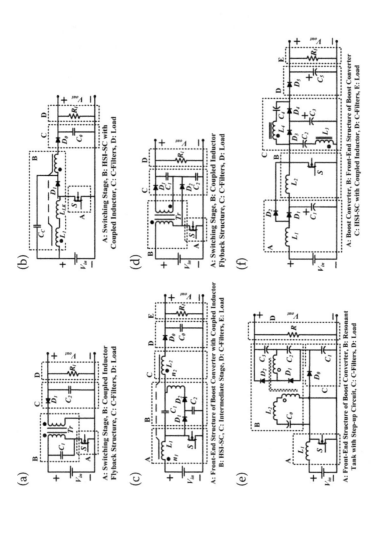

FIGURE 3.21 Power Circuits: (a) Standard Ground-Coupled Inductor-Based Step-Up Flying Converter Configuration, (b) Coupled Inductor-Based Converter Configuration, (c) High Step-Up Switched-Capacitor and Coupled Inductor-Based DC-DC Converter Configuration, (d) Hybrid-Flyback-Boost Converter Configuration, (e) Boost Converter Configuration with Integrated Transformer, (f) Coupled Inductor-Based Quadratic Boost Converter.

ratio, coupling factor, and turn-on time of the switch. To design this converter configuration, a conventional flyback converter front-end structure and C-filter are used with common grounding.

$$G_{C-\text{flying converter}} = \frac{V_{out}}{V_{in}} = \frac{-(N_2/N_1)\Delta}{1 - \Delta} = \frac{T_{on}N}{T_{on} - T} = \frac{T_{on}f_s N}{T_{on}f_s - 1} \qquad (3.35)$$

To attain a high voltage conversion ratio, we use the coupled inductor as a transformer at the load side of the converter configurations—the power circuit of this converter configuration is depicted in Figure 3.21(b) [157]. The proper duty cycle and turn ratio are selected, to achieve high voltage conversion. Switching stage, a coupled inductor with HSI-SC, and C-filter are the three stages used to design the converter configurations. To reduce the converter size, it is possible to mount both coupled inductors on a single magnetic core.

The power circuit of the high step-up switched-capacitor and coupled inductor-based DC-DC converter configuration is shown in Figure 3.21(c) [156–158]. Coupled inductor-based FES-BC, intermediate stage, HIS-SC stage, and C-filter are the four stages used to design this configuration. In this configuration, the coupled inductor is used to design the intermediate stage and high-voltage conversion ratio achieved by adjusting the turn-on time and turns ratio. This circuit assists in energy recycling for active clamping, but the disadvantages are the high voltage across the switch and losses occurring due to leakage of inducing energy.

The power circuit of a hybrid-flyback-boost converter configuration is shown in Figure 3.21(d) [159]. A switching stage, flyback converter front-end structure and two C-filter stages are used to design this configuration. To reduce the voltage across the switch and to achieve a high-voltage conversion ratio, the output of the flyback and boost converter are connected in series. The power circuit of a boost converter configuration with an integrated transformer is shown in Figure 3.21(e) [160]. In this circuit, to avoid a high current ripple at the input side, the auxiliary circuit is designed by the capacitor and integrated transformer. The configuration is designed by connecting an auxiliary circuit with a boost converter. To achieve a high-voltage conversion ratio, the voltage doubler is used in the auxiliary circuit. The transformer and capacitor form a resonant tank, and thus a quasi-resonant mode is possible. A resonant tank, FES-BC, and C-filter stages are used to design this converter configuration.

The turn ratio of the transformer is adjusted to attain a high voltage at the output side. The power circuit of a coupled inductor is based on a quadratic boost converter in Figure 3.21(f) [161]. The converter is designed by combining the feature of a conventional quadratic boost, coupled with an inductor-based HIS-SC to achieve a high-voltage conversion ratio. This configuration requires four stages: a quadratic boost converter is designed by using the initial two stages, a coupled inductor-based HSI-SC is the third stage, and a filter is the fourth stage.

The power circuit of a lossless clamping tapped buck converter (LC-TBC) configuration is shown in Figure 3.22(a) [162]. The voltage spikes across the switch are recovered by using a lossless clamp circuit. A buck converter and HSI-SC

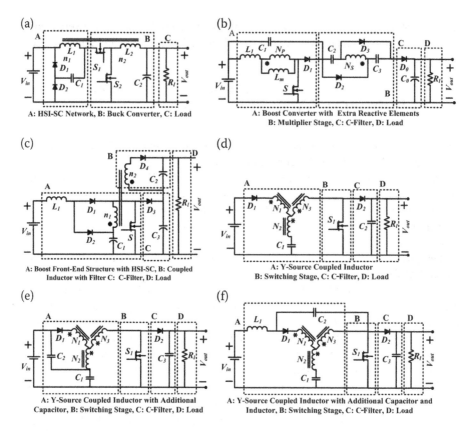

FIGURE 3.22 Power Circuits: (a) Lossless Clamping Tapped Buck Converter (LC-TBC) Configuration, (b) Multiplier-Based High Step-Up Converter Configuration, (c) High Step-Up Coupled Inductor-Based Quadratic Boost Converter Configuration, (d) Y-Source DC-DC Converter Configuration, (e) Y-Source DC-DC Converter Configuration with an Additional Capacitor, (f) Y-Source DC-DC Converter Configuration with Additional Capacitor and Inductor.

stages are used to design a LC-TBC configuration where an inductor of a buck converter and a HSI-SC is commonly coupled which each other. The voltage conversion ratio of LC-TBC configuration is provided in (3.36).

$$G_{LC-TBC} = \frac{V_{out}}{V_{in}} = \frac{\Delta}{\Delta + N(1 - \Delta)} = \frac{T_{on}f_s}{(1 - T_{on}f_s)N + T_{on}f_s} \quad (3.36)$$

The power circuit of a multiplier-based high step-up converter configuration is shown in Figure 3.22(b) [163]. In this configuration, parallel charging and series discharging of two capacitors are taking place with inductor-stored energy to attain a high-voltage conversion ratio. Coupled inductor leakage energy recycled by using a clamp circuit. A reactive element with the boost converter stage is used to design the first stage of the converter. The second stage of the converter is designed by a coupled inductor-based multiplier and the third stage is designed by the filter. The low-resistance control switch is used to minimize the on-state conduction losses.

The power circuit of a high step-up coupled inductor is based on a quadratic boost converter configuration shown in Figure 3.22(c) [164]. The converter is designed to achieve a high-voltage conversion ratio with a low-voltage switch stress by utilizing FES-BC, a coupled inductor-based boost converter, and a C-filter. The voltage conversion ratio of the configuration is raised by changing the coupling factor of coupled inductors.

The power circuit of the Y-source DC-DC converter configuration is shown in Figure 3.22(d) [164–167]. A Y-source configuration is designed to achieve a high-voltage conversion ratio with a lesser duty cycle by utilizing three stages: switching stage, Y-source impedance network, and C-filter. The converter configuration has all the advantages of a Y-source, but the pulsating and discontinuous nature of current is the major drawback of the converter.

The power circuit of the Y-source DC-DC converter configuration with the additional capacitor is shown in Figure 3.22(e) [164–167]. A simple modification is done by adding a capacitor in a Y-source converter to remove drawback of discontinuous current and to achieve smooth current. Continuous current is only achieved by selecting the appropriate capacitor. The converter is designed to achieve a high-voltage conversion ratio by utilizing three stages: switching stage, a Y-source with parallel capacitor, and C-filter. A high inrush current is the major drawback of this converter due to reactance at the input side.

The power circuit of the Y-source DC-DC converter configuration with an additional capacitor and inductor is shown in Figure 3.22(f) [164–167]. A simple modification is done by adding an inductor and capacitor in a Y-source converter to smoothen current. The converter is designed to achieve a high-voltage conversion ratio by utilizing three stages: switching stage, a Y-source with an inductor and capacitor, and C-filter. Regrettably, in this configuration, the voltage across devices is high.

The power circuit of the quasi Y-source DC-DC converter configuration is shown in Figure 3.23(a) [164–167]. We design the configuration to achieve a high-voltage conversion ratio and to achieve a continuous, smooth current by utilizing three stages: switching stage, quasi Y-source, and C-filter. The DC current is blocked through a coupled inductor by using two capacitors. Thus, the core is prevented from saturation.

The power circuit of the tapped coupled inductor is based on the buck converter configuration shown in Figure 3.23(b) [168,169]. We combine tapped coupled inductor and buck converter features to design a converter configuration, and voltage at the output is adjusted by changing the inductor tapping. For a coupled inductor leakage inductance, a high spike in voltage across a switch is observed, but conduction loss and switching losses are reduced. The voltage conversion ratio is adjusted by changing the tapping of the inductors.

The power circuit of the active clamp tapped buck converter configuration is shown in Figure 3.23(c) [170,171]. Using this configuration, the voltage spike problem is reduced by utilizing an active clamp circuit. The configuration is

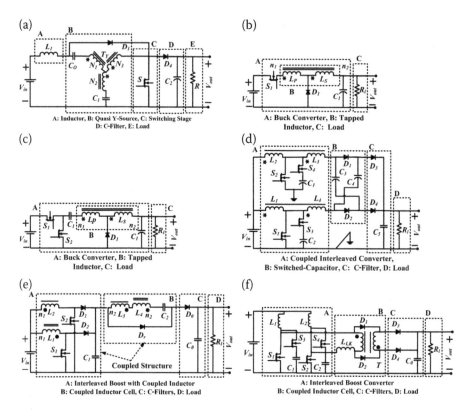

FIGURE 3.23 Power Circuits: (a) Quasi Y-Source DC-DC Converter Configuration, (b) Tapped Coupled Inductor-Based Buck Converter Configuration, (c) Active Clamp Tapped Buck Converter Configuration, (d) Switched Capacitor and Coupled Inductor-Based Interleaved Boost Converter Configuration, (e) Multiplier-Based Interleaved Boost Converter Configuration, (f) Built-in-Transformer-Based Interleaved Boost Converter Configuration.

designed to achieve a high step-down ratio by combining the feature of the active clamp tapped inductor and buck converter.

When the power circuit of the switched capacitor and coupled inductor base is interleaved, it shows a boost converter configuration in Figure 3.23(d) [172,173]. In this configuration, the voltage across the switch is reduced and a high-voltage conversion ratio is achieved by employing a switched capacitor. Zero-voltage transition (ZVT) can be used throughout the switching cycle to reduce switching losses. We design the converter configuration in three stages: switched-capacitor, interleaved boost converter, and C-filter.

The power circuit of the multiplier-based interleaved boost converter is shown in Figure 3.23(e) [174]. The high-voltage conversion ratio can be achieved by adjusting the duty cycle and a coupling factor of inductors. Switches with a low-voltage rating and small size inductors are fit for this configuration for sharing current capability at the input port. Zero-current technique (ZCT) reduces the EMI and switching losses and thus, increases the performance of the converter. The converter is designed by utilizing three stages: coupled inductor-based voltage multiplier, coupled inductor-based interleaved converter, and C-filter.

The power circuit of a built-in-transformer-based interleaved boost converter configuration is shown in Figure 3.23(f) [175]. The bridge rectifier is used and leakage of the transformer is used for the current recovery of the bridge. The converter is highly efficient for soft switching and the configuration is designed by utilizing three stages: intermediate rectifier designed by the coupled inductor, interleaved boost converter, and filter.

The power circuit of the coupled inductor interleaved winding boost converter configuration is shown in Figure 3.24(a) [176]. To restrict the voltage across a switch, sufficient, clamping circuitry is used and switching losses are reduced by ZCS. The voltage conversion ratio is changed by adjusting the coupled inductor turns ratio. The converter is designed by using three stages: coupled inductor circuitry, coupled inductor-based interleaved converter, and C-filter.

The power circuit of a voltage doubler based built-in-transformer interleaved boost converter is shown in Figure 3.24(b) [174]. A voltage double-based built-in

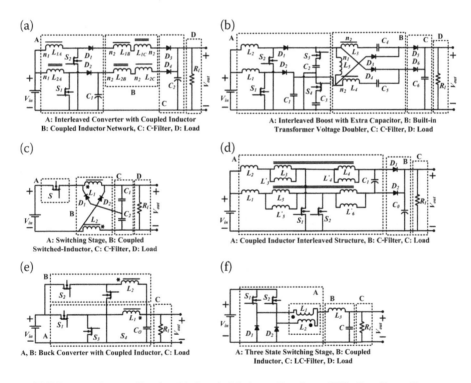

FIGURE 3.24 Power Circuits: (a) Coupled Inductor Interleaved Winding Boost Converter Configuration, (b) Voltage Doubler Based Built-in-Transformer Interleaved Boost Converter, (c) Coupled Switched-Inductor-Based Buck Converter Configuration, (d) Coupled Inductor-Based Interleaved Boost Converter Configuration with an Intermediate Capacitor, (e) Coupled Inductor-Based Interleaved Buck Converter Configuration, (f) Three Switching State Cell-Based Buck Converter Configuration.

transformer is designed by using two capacitors, two diodes, and three windings. Parallel charging and series discharging operations of the capacitor are performed to double the voltage. We can use an active clamp technique to recover the leakage energy. The converter configuration is designed by using three stages: voltage doubler based built-in-transformer, interleaved boost converter with extra capacitor, and C-filter.

The power circuit of a coupled switched-inductor-based buck converter configuration is shown in Figure 3.24(c) [177]. In this configuration, it requires no active clamp circuitry to recover leakage energy—the configuration is designed by replacing a filter inductor of a buck converter with a coupled inductor. The configuration combines the feature in three stages: switched coupled inductor, switching stage, and C-filter. The power circuit of a coupled inductor-based interleaved boost converter configuration with the intermediate capacitor is shown in Figure 3.24(d) [178]. In this configuration, the intermediate capacitor is used to make a converter efficient and to reduce converter-switching losses. The power circuit of a coupled inductor-based interleaved buck converter configuration is shown in Figure 3.24(e) [179–182]. Less current ripple in the output is achieved by selecting a proper coupling inductor. The power circuit of three switching state cell-based buck converter configurations is shown in Figure 3.24(f) [183,184]. The peak of active switch current is reduced by employing three switching state cells. The configuration also reduces the current ripple and design by using two stages: three switching state cells and LC-filter. Two coupled inductors, two switches, and two diodes are required to design a three-switching state cell.

3.8   LUO DC-DC CONVERTER FAMILY

Luo DC-DC converters are well-accepted power converter configurations that provide an excellent source to achieve a high-voltage conversion ratio. Self-lift, re-lift, and triple lift are voltage-boosting techniques used in Luo converters to remove the effect of parasitic components and to achieve high output voltage at the output terminal of the converter with higher efficiency [185,189]. Switched capacitor (SC) and switched inductor (SI) concepts are used to design circuitry for voltage-lifting purposes.

The power circuit of a re-lift positive output Luo converter (RL-PO-LC) configuration is shown in Figure 3.25(a) [185,186]. This converter configuration gives positive output at the output terminal, designed with the help of re-lift circuitry and two control switches. The features of the hybrid combination of a SI and the SC are combined to design a voltage re-lift circuitry. Re-lift circuitry, FES-BBC, and LC-filter are three stages required to design a RL-PO-LC configuration and the voltage conversion ratio given in (3.37).

$$G_{RL-PO-LC} = \frac{V_{out}}{V_{in}} = \frac{2}{1-\Delta} = \frac{2}{1-\frac{T_{on}}{T}} = \frac{2}{1-T_{on}f_s} \qquad (3.37)$$

The power circuit of a triple-lift positive output Luo converter (TL-PO-LC) configuration is shown in Figure 3.25(b) [185–187]. This converter configuration gives

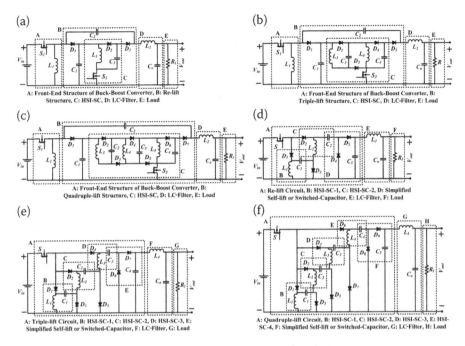

FIGURE 3.25 Power Circuits: (a) Re-Lift Positive Output Luo Converter (RL-PO-LC) Configuration, (b) Triple-Lift Positive Output Luo Converter (TL-PO-LC) Configuration, (c) Quadruple-Lift Positive Output Luo Converter (QL-PO-LC) Configuration, (d) Simplified Re-Lift Positive Output Luo Converter (S-RL-PO-LC) Configuration, (e) Simplified Triple-Lift Positive Output Luo Converter (S-TL-PO-LC) Configuration, (f) Simplified Quadruple-Lift Positive Output Luo Converter (S-QL-PO-LC) Configuration.

a positive output at the output terminal, designed with the help of triple-lift circuitry and two control switches. The features of the hybrid combination of a SI (double-switched inductor configuration) and SC are combined to design a voltage triple-lift circuitry. Triple-lift circuitry, FES-BBC, and LC-filter are three stages required to design a TL-PO-LC configuration and the voltage conversion ratio given in (3.38).

$$G_{\text{TL-PO-LC}} = \frac{V_{out}}{V_{in}} = \frac{3}{1-\Delta} = \frac{3}{1-\frac{T_{on}}{T}} = \frac{3}{1-T_{on}f_s} \tag{3.38}$$

The power circuit of a quadruple-lift positive output Luo converter (QL-PO-LC) configuration is shown in Figure 3.25(c) [185–187]. This converter configuration gives a positive output at the output terminal, designed with the help of a quadruple-lift circuitry and two control switches. The feature of the hybrid combination of a SI (triple-switched inductor configuration) and SC are combined to design a voltage quadruple-lift circuitry. Quadruple-lift circuitry, FES-BBC, and LC-filter are three stages required to design a TL-PO-LC configuration and the voltage conversion ratio given in (3.39).

$$G_{QL-PO-LC} = \frac{V_{out}}{V_{in}} = \frac{4}{1 - \Delta} = \frac{4}{1 - \frac{T_{on}}{T}} = \frac{4}{1 - T_{on}f_s} \tag{3.39}$$

The power circuit of a simplified re-lift positive output Luo converter (S-RL-PO-LC) configuration is shown in Figure 3.25(d) [185–187]. This converter configuration gives a positive output at the output terminal, designed with the help of a simplified re-lift circuitry and a single control switch. The features of two hybrid combinations of a SI and a SC (HSI-SC-1, HSI-SC-2) and simplified self-lift circuitry or switched capacitor are combined to design a simplified re-lift circuitry. A LC-filter and simplified re-lift circuitry are two stages required to design a S-RL-PO-LC configuration and the voltage conversion ratio is given in (3.40).

$$G_{S-RL-PO-LC} = \frac{V_{out}}{V_{in}} = \frac{2}{1 - \Delta} = \frac{2}{1 - \frac{T_{on}}{T}} = \frac{2}{1 - T_{on}f_s} \tag{3.40}$$

The power circuit of a simplified triple-lift positive output Luo converter (S-TL-PO-LC) configuration is shown in Figure 3.25(e) [185–187]. This converter configuration gives a positive output at the output terminal, designed with the help of a simplified triple-lift circuitry and a single control switch. The feature of three hybrid combinations of a SI and SC (HSI-SC-1, HSI-SC-2, HSI-SC-3) and simplified self-lift circuitry or switched capacitor are combined to design a simplified triple-lift circuitry. A LC-filter and simplified triple-lift circuitry are two stages required to design a S-RL-PO-LC configuration and the voltage conversion ratio is given in (3.41).

$$G_{S-TL-PO-LC} = \frac{V_{out}}{V_{in}} = \frac{3}{1 - \Delta} = \frac{3}{1 - \frac{T_{on}}{T}} = \frac{3}{1 - T_{on}f_s} \tag{3.41}$$

The power circuit of a simplified quadruple-lift positive output Luo converter (S-QL-PO-LC) configuration is shown in Figure 3.25(f) [185–187]. This converter configuration gives positive output at the output terminal, designed with the help of a simplified quadruple-lift circuitry and single control switch. The features of four hybrid combinations of a SI and SC (HSI-SC-1, HSI-SC-2, HSI-SC-3, HSI-SC-4) and simplified self-lift circuitry or switched capacitor are combined to design a simplified quadruple-lift circuitry. A LC-filter and simplified quadruple-lift circuitry are two stages required to design a S-RL-PO-LC configuration and the voltage conversion ratio is given in (3.42).

$$G_{S-QL-PO-LC} = \frac{V_{out}}{V_{in}} = \frac{4}{1 - \Delta} = \frac{4}{1 - \frac{T_{on}}{T}} = \frac{4}{1 - T_{on}f_s} \tag{3.42}$$

The power circuit of a simplified re-lift negative output Luo converter (S-RL-NO-LC) configuration is shown in Figure 3.26(a) [187–189]. The feature of two hybrid

(a)

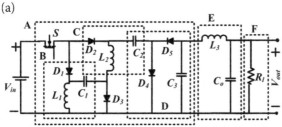

A: Re-lift Circuit, B: HSI-SC-1, C: HSI-SC-2, D: Negative
Simplified Self-lift or Switched-Capacitor, E: LC-Filter, F: Load

(b)

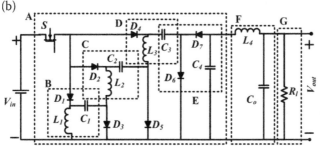

A: Triple-lift Circuit, B: HSI-SC-1, C: HSI-SC-2, D: HSI-SC-3, E: Negative
Simplified Self-lift or Switched-Capacitor, F: LC-Filter, G: Load

(c)

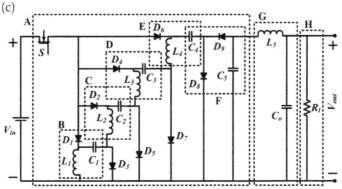

A: Quadruple-lift Circuit, B: HSI-SC-1, C: HSI-SC-2, D: HSI-SC-3, E: HSI-SC-4,
F: Negative Simplified Self-lift or Switched-Capacitor, G: LC-Filter, H: Load

FIGURE 3.26 Power Circuits: (a) Simplified Re-Lift Negative Output Luo Converter (S-RL-NO-LC) Configuration, (b) Simplified Triple-Lift Negative Output Luo Converter (S-RL-NO-LC) Configuration, (c) Simplified Triple-Lift Negative Output Luo Converter (S-RL-NO-LC) Configuration.

combinations of a SI and SC (HSI-SC-1, HSI-SC-2) and a negative simplified self-lift circuitry or switched capacitor are combined to design a simplified negative re-lift circuitry. A LC-filter and a simplified negative re-lift circuitry are two stages required to design a S-RL-NO-LC configuration and the voltage conversion ratio is given in (3.43).

$$G_{\text{S}-\text{RL}-\text{NO}-\text{LC}} = \frac{V_{out}}{V_{in}} = \frac{-2}{1-\Delta} = \frac{-2}{1-\frac{T_{on}}{T}} = \frac{-2}{1-T_{on}f_s} \qquad (3.43)$$

The power circuit of a simplified triple-lift negative output Luo converter (S-TL-NO-LC) configuration is shown in Figure 3.26(b) [185–189]. The features of three hybrid combinations of a SI and SC (HSI-SC-1, HSI-SC-2, HSI-SC-3) and a negative simplified self-lift circuitry or SC are combined to design a simplified negative triple-lift circuitry. A LC-filter and a simplified negative triple-lift circuitry are two stages required to design a S-TL-NO-LC configuration and the voltage conversion ratio is given in (3.44).

$$G_{\text{S}-\text{TL}-\text{NO}-\text{LC}} = \frac{V_{out}}{V_{in}} = \frac{-3}{1-\Delta} = \frac{-3}{1-\frac{T_{on}}{T}} = \frac{-3}{1-T_{on}f_s} \qquad (3.44)$$

The power circuit of a simplified quadruple-lift negative output Luo converter (S-QL-NO-LC) configuration is shown in Figure 3.26(c) [185–189]. The features of four hybrid combinations of a SI and SC (HSI-SC-1, HSI-SC-2, HSI-SC-3, HSI-SC-4) and a negative simplified self-lift circuitry or SC are combined to design a simplified negative quadruple-lift circuitry. A LC-filter and a simplified negative quadruple-lift circuitry are two stages required to design a S-QL-NO-LC configuration and the voltage conversion ratio is given in (3.45).

$$G_{\text{S}-\text{QL}-\text{NO}-\text{LC}} = \frac{V_{out}}{V_{in}} = \frac{-4}{1-\Delta} = \frac{-4}{1-\frac{T_{on}}{T}} = \frac{-4}{1-T_{on}f_s} \qquad (3.45)$$

3.9 Z-SOURCE DC-DC CONVERTER CONFIGURATIONS

In 2020, Peng introduced the concept of a Z-source network. The Z-source network is mainly used to achieve a high DC or AC step-up output voltage from the power converter application [190–194]. In a DC-DC power converter, the primary advantage of the Z-source network is its high-voltage boosting capability. Therefore, the Z-source converter can be an excellent choice for high-voltage renewable energy applications [190–196].

The high-gain DC-DC converter configuration is based on conventional a Z-source network and is shown in Figure 3.27 [190–196].

The power circuitry is designed by using four stages: diode stage, Z-source network, switching stage, and LC-filter. The load is connected across the capacitor of the LC stage. The voltage gains of the converter (shown in Figure 3.27) are given in (3.46)

$$G_{\text{Z}-Source} = \frac{V_{out}}{V_{in}} = \frac{1-\Delta}{1-2\Delta} = \frac{1-\frac{T_{on}}{T}}{1-2\frac{T_{on}}{T}} = \frac{1-T_{on}f_s}{1-2T_{on}f_s} \qquad (3.46)$$

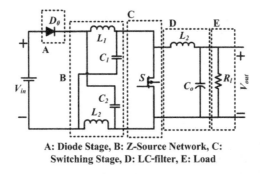

A: Diode Stage, B: Z-Source Network, C:
Switching Stage, D: LC-filter, E: Load

FIGURE 3.27 High-Gain DC-DC Converter Configuration Based on the Conventional Z-Source Network.

where the value of the duty cycle (Δ) is smaller than 0.5. The significant advantages of the converter (shown in Figure 3.27) are as follows:

1. There is lower voltage stress across semiconductor devices.
2. There is no diode connected at the output side of the converter. Therefore, the reverse recovery problem of the output-side diode is eliminated.
3. The converter provides a high boosting factor without using a transformer with a high turn ratio.
4. The controlled switch is turned on with a duty cycle lower than 0.5. Therefore, the conduction loss of the controlled switch is reduced.

In a conventional Z-source network-based DC-DC converter, the input and output terminals are not at the common ground, which is the main drawback of circuit. The quasi Z-source circuit provides a solution for the DC-DC converter to overcome the grounding problem, when compared to conventional Z source circuit. Several power circuits with quasi Z-source are well elaborated in the literatures and for its practical applications [193–196].

The high DC-DC converter is based on the conventional quasi Z-source network shown in Figure 3.28. The power circuitry is designed by using three stages: quasi Z-source network, switching stage, and C-filter. The voltage gains of the high

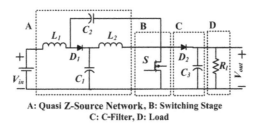

A: Quasi Z-Source Network, B: Switching Stage
C: C-Filter, D: Load

FIGURE 3.28 High DC-DC Converter Based on the Conventional Quasi Z-Source Network.

DC-DC converter are based on the conventional quasi Z-source network, shown in Figure 3.28, and given in (3.47)

$$G_{Quasi-Source} = \frac{V_{out}}{V_{in}} = \frac{1}{1 - 2\Delta} = \frac{1}{1 - 2\frac{T_{on}}{T}} = \frac{1}{1 - 2T_{on}f_s} \quad (3.47)$$

where the value of the duty cycle (Δ) is smaller than 0.5.

To improve the voltage boosting capability of the quasi Z-source-based converter, an additional quasi Z-source or conventional Z-source network is adopted in the power circuitry. The high-gain DC-DC converter is based on the modified two quasi Z-source network shown in Figure 3.29 [195,196]. This converter (Figure 3.29) combines the functionality of a two quasi Z-source network to achieve high output voltage. The power circuitry is designed by using three stages: modified two quasi Z-source network, switching stage, and C-filter.

The voltage gains of the high-gain DC-DC converter are based on a modified two quasi Z-source network, shown in Figure 3.29, and given in (3.48)

$$G_{Two-Quasi-Source} = \frac{V_{out}}{V_{in}} = \frac{1}{1 - 3\Delta} = \frac{1}{1 - 3\frac{T_{on}}{T}} = \frac{1}{1 - 3T_{on}f_s} \quad (3.48)$$

where the value of the duty cycle (Δ) is smaller than 0.5 and the high-gain DC-DC converter is based on the modified three quasi Z-source network shown in Figure 3.30 [195,196].

The converter shown in Figure 3.30 combines the functionality of a three quasi Z-source network to achieve high output voltage. The power circuitry is designed by using three stages: modified three quasi Z-source network, switching stage, and C-filter. The voltage gains of a high-gain DC-DC converter are based on the modified three quasi Z-source network, shown in Figure 3.30, and given in (3.49)

$$G_{Three-Quasi-Source} = \frac{V_{out}}{V_{in}} = \frac{1}{1 - 4\Delta} = \frac{1}{1 - 4\frac{T_{on}}{T}} = \frac{1}{1 - 4T_{on}f_s} \quad (3.49)$$

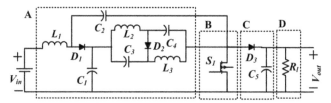

A: Modified Two Quasi Z-Source Network, B: Switching Stage, C: C-Filter, D: Load

FIGURE 3.29 High-Gain DC-DC Converter Based on Modified Two Quasi Z-Source Network.

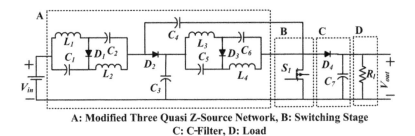

A: Modified Three Quasi Z-Source Network, B: Switching Stage
C: C-Filter, D: Load

FIGURE 3.30   High-Gain DC-DC Converter Based on the Modified Three Quasi Z-Source
Network.

where the value of the duty cycle (Δ) is smaller than 0.5 and the high-gain DC-DC
converter is based on a hybrid combination of two quasi and conventional Z-source
networks are shown in Figure 3.31 [195,196]. The converter shown in Figure 3.31
combines the functionality of a quasi Z-source network and conventional Z-source
network to achieve a high-output voltage. The power circuitry is designed by using
five stages: diode stage, a hybrid combination of quasi and conventional Z-source
networks, switching stage, C-filter and load. The voltage gains of the high-gain
DC-DC converter are based on a hybrid combination of two quasi and conventional
Z-source networks, shown in Figure 3.31, and given in equation (3.50),

$$G_{Three-Quasi-Source} = \frac{V_{out}}{V_{in}} = \frac{1}{1 - 4\Delta} = \frac{1}{1 - 4\frac{T_{on}}{T}} = \frac{1}{1 - 4T_{on}f_s} \qquad (3.50)$$

where the value of the duty cycle (Δ) is smaller than 0.5.
   The common advantages of the quasi Z-source-based DC-DC converter are as
follows:

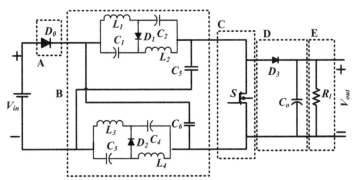

A: Diode Stage, B: Hybrid Combination of Quasi and Conventional
Z-Source Networks, C: Switching Stage, D: C-Filter, E: Load

FIGURE 3.31   High-Gain DC-DC Converter Based on a Hybrid Combination of Quasi and
Conventional Z-Source Networks.

- There is lower voltage stress across semiconductor devices.
- There is continuous input current.
- In most of the cases, there is common ground between the input and output terminal.
- The converter provides a high boosting factor without using a transformer with a high turn ratio.
- We turn the controlled switch on with a duty cycle lower than 0.5. Therefore, conduction losses of the controlled switch are reduced.

3.10 COCKCROFT WALTON VOLTAGE MULTIPLIER-BASED MULTILEVEL DC-DC CONVERTER FAMILY

In the last few years, several DC-DC converters were achieved by combining the features of the conventional DC-DC converter and Cockcroft Walton voltage multiplier to achieve high voltage at the output side [197–200]. These configurations are called multilevel DC-DC converter configurations and have the following advantages:

- The voltage across semiconductor devices is low.
- High voltage is achieved at an average duty cycle.
- The number of levels of the converter can easily be increased by one level by adding only two diodes and two capacitors without disturbing the main boosting stage.
- The configurations are modular and low in weight.
- They easily fit for DC-DC-AC multilevel applications.
- The required number of switches is less.

The power circuit of a hybrid voltage multiplier-based N-stage DC-DC converter configuration is shown in Figure 3.32(a) [71,197–200]. This configuration is designed by attaching an N-stage voltage multiplier with a conventional boost converter. It requires two capacitors and two diodes to design a one-stage voltage multiplier. The output voltage of the converter can be increased in two ways: one way is by increasing the duty cycle and another is by increasing the number of stages of the voltage multiplier. The voltage across the switch is low, and the number of stages can quickly increase by connecting two diodes' and two capacitors' circuitry at the output side. A N-stage voltage multiplier, FES-BC, and C-filter are the three stages used to designed a hybrid voltage multiplier-based N-stage DC-DC converter configuration. This configuration gives non-inverting output voltage, and the output voltage is dependent upon the number of multiplier stages and duty cycle. The voltage conversion ratio of a hybrid voltage multiplier-based N-stage DC-DC converter configuration is given in (3.51).

$$G_{N-stage\ converter} = \frac{V_{out}}{V_{in}} = \frac{1-N}{\Delta-1} = \frac{(1-N)}{\frac{T_{on}}{T}-1} = \frac{N-1}{1-T_{on}f_s} \qquad (3.51)$$

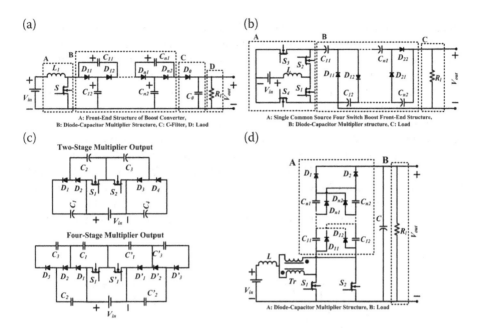

FIGURE 3.32 Power Circuits: (a) Hybrid Voltage Multiplier-Based N-Stage DC-DC Converter Configuration, (b) DC-DC Converter Configuration with N-Stage Cockcroft Walton (CW) Voltage Multiplier, (c) Voltage Multiplier with Two Switches DC-DC Converter Configuration, (d) Transformer-Multiplier Three-State Switching Cell Hybrid Converter Configuration.

The power circuit of a DC-DC converter configuration with a N-stage Cockcroft Walton (CW) voltage multiplier is shown in Figure 3.32(b) [71,198–200]. A N-stage CW voltage multiplier, front-end structure of single-input four switches boost converter, and C-filter are three sections used to design this converter configuration. Thus, this configuration combines the feature of a boost converter with four switches and a N-stage CW voltage multiplier. Without utilizing coupled inductors and transformer, this configuration provides a high-output voltage at the average duty cycle. This configuration gives a non-inverting output voltage and the output voltage depends on the number of multiplier stages and duty cycle. The voltage conversion ratio of a DC-DC converter configuration with a N-stage Cockcroft Walton (CW) voltage multiplier is provided in (3.47).

$$G_{N-stage-CW-Converter} = \frac{V_{out}}{V_{in}} = \frac{N}{1 - \Delta} = \frac{N}{1 - \frac{T_{on}}{T}} = \frac{N}{1 - T_{on}f_s} \qquad (3.52)$$

The power circuit of a voltage multiplier with two switches DC-DC converter configuration is shown in Figure 3.32(c) [71,201]. The two switches are operating alternately to achieve high voltage at the output side. This configuration weighs less because we do not require the inductor. The C-filter stage is avoided because this configuration has capacitors in series at the output side. The voltage conversion ratio depends on the number of stages voltage multiplier or number of capacitors. The power circuit of a transformer-multiplier three-state switching cell hybrid converter configuration is shown in Figure 3.32(d) [71,202]. The current flow through the switch is half of the output current, and high frequency can be used to decrease the voltage ripple, current ripple, volume, and size of the converter. The voltage conversion ratio is raised by increasing the multiplier levels at the output side.

The power circuit of a multilevel boost converter (MBC) is shown in Figure 3.33(a) [31,71,203]. The converter combines the feature of a conventional boost converter and a CW voltage multiplier. The power circuit of a multilevel buck-boost converter (MBBC) is shown in Figure 3.33(b) [71,203–205]. The converter combines the features of the conventional buck-boost converter and CW

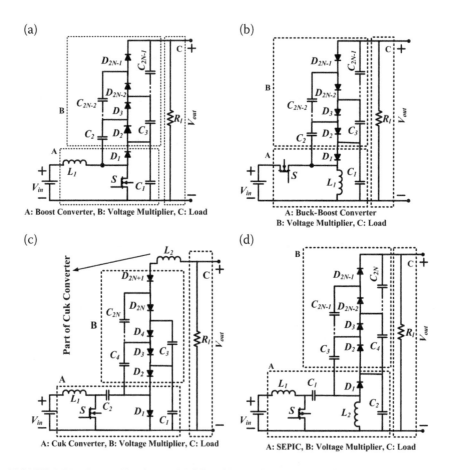

FIGURE 3.33 Power Circuits: (a) Multilevel Boost Converter (MBC), (b) Multilevel Buck-Boost Converter (MBBC), (c) Multilevel Cuk Converter (MCC), and Multilevel SEPIC Converter (MSC).

voltage multiplier. The power circuit of a multilevel Cuk converter (MCC) is shown in Figure 3.33(c) [206]. The converter combines the feature of a Cuk converter and CW voltage multiplier. The power circuit of a multilevel SEPIC converter (MSC) is shown in Figure 3.33(d) [207–209]. The converter combines the features of a SEPIC converter and a CW voltage multiplier. The MBC, MBBC, MCC, and MSC configurations provide a higher-voltage conversion ratio with the low voltage across a switch compared to the conventional boost converter [203–209]. The following are the major advantages of MBC, MBBC, MCC, and MSC:

- High voltage is achieved at the output of the converters without using the coupled inductor and transformer.
- The voltage across a switch and diodes is low compared to the output voltage.
- The configuration is easily extendable by adding an additional level of the multiplier.
- It requires single switch to design converter configurations.
- It is a self-voltage balanced structure.

The MBC and MSC provide a high non-inverting output voltage and the output voltage depends on the number of levels and duty cycle. The MBBC and MCC provide a high inverting output voltage and the output voltage depends on the number of levels and duty cycle. The voltage conversion ratios of MBC, MBBC, MCC, and MSC are given in equations (3.53)–(3.56), respectively, and where N is several levels of the converter.

$$G_{MBC} = \frac{V_{out}}{V_{in}} = \frac{N}{1-\Delta} = \frac{N}{1-\frac{T_{on}}{T}} = \frac{N}{1-T_{on}f_s} \tag{3.53}$$

$$G_{MBC} = \frac{V_{out}}{V_{in}} = \frac{N}{1-\Delta} = \frac{N}{1-\frac{T_{on}}{T}} = \frac{N}{1-T_{on}f_s} \tag{3.54}$$

$$G_{MCC} = \frac{V_{out}}{V_{in}} = \frac{-1-(N-1)\Delta}{1-\Delta} = \frac{-1-(N-1)T_{on}f_s}{1-T_{on}f_s} \tag{3.55}$$

$$G_{MSC} = \frac{V_{out}}{V_{in}} = \frac{(N-1)\Delta+1}{1-\Delta} = \frac{(N-1)T_{on}f+1_s}{1-T_{on}f_s} \tag{3.56}$$

The power circuit of a switched-inductor-based multilevel boost converter (switched inductor MBC) configuration is shown in Figure 3.34 [71,210,211]. The configuration combines the features of the conventional boost converter, voltage multiplier, and switched inductor. A higher-voltage conversion ratio is possible compared to MBC. A switched inductor concept is used to achieve

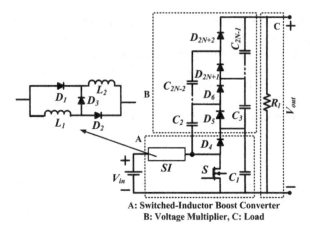

A: Switched-Inductor Boost Converter
B: Voltage Multiplier, C: Load

FIGURE 3.34 Power Circuit of Switched-Inductor-Based Multilevel Boost Converter (Switched Inductor MBC) Configuration.

higher voltage by minimizing the number of components and number of levels of the multiplier.

The voltage conversion ratio is given in (3.57) which is depends on the number of levels of multiplier and duty cycles.

$$G_{Switch\ Inductor\ MBC} = \frac{V_{out}}{V_{in}} = \frac{1+\Delta}{1-\Delta}N = \frac{1+\frac{T_{on}}{T}}{1-\frac{T_{on}}{T}} = \frac{1+T_{on}f_s}{1-T_{on}f_s}N \qquad (3.57)$$

Nowadays, many multilevel DC-DC converter configurations are achieved without an inductor and transformer to make circuitry low and low weight. A transformer-less and inductor-less (without magnetic element) DC-DC multilevel converter configuration is briefly explained in Chapter 4.

3.11 COMPARISON OF DC-DC MULTISTAGE CONVERTERS

To choose a suitable boost or buck converter from the previously mentioned configurations, it is essential to compare DC-DC converters in terms of voltage gain, several capacitors, inductors, diodes, switches, switch voltage, and transformers. The DC-DC converters of the multistage SCBPC family, multistage SIBPC family, coupled inductor or transformer-based converter family, Luo DC-DC converter family, Z-source converter family Cockcroft Walton voltage multiplier-based multilevel DC-DC converter family are compared in Tables 3.1 to 3.6 [71]. In Table 3.7, converters are compared in terms of cost, efficiency, reliability, advantages, and disadvantages [71].

TABLE 3.1
Comparison of Multistage SCBPC Family

Figure No.	Switch Voltage Stress	NL	NC	NS	ND	NT
3.13(a)	$V_{in}/(1 - \Delta)$	1	2	1	2	–
3.13(b)	$V_{in}/(1 - \Delta)$	1	4	1	4	–
3.13(c)	$V_{in}/(1 - \Delta)$	2	3	1	2	–
3.13(d)	$V_{in}/(1 - \Delta)$	2	3	1	2	–
3.13(e)	$V_{in}/(1 - \Delta)$	2	3	1	2	–
3.13(f)	$V_{in}/(1 - \Delta)$	2	3	1	2	–
3.14(a)	$V_{in}/(1 - \Delta)$	2	4	1	3	–
3.14(b)	$V_{in}/(1 - \Delta)$	1	3	1	3	–
3.14(c)	$V_{in}/(1 - \Delta)$	2	4	1	3	–
3.14(d)	$V_{in}/(1 - \Delta)$	1	3	1	3	–
3.14(e)	$V_{in}/(1 - \Delta)$	1	3	1	3	–
3.14(f)	$V_{in}/(1 - \Delta)$	2	3	1	2	–
3.15(a)	Max. V_{out}	1	3	4	5	–
3.16(a)	$V_{in}/(1 - \Delta)^2$	2	4	1	5	–
3.16(b)	$V_{in}/(1 - \Delta)^2$	3	4	1	4	–
3.16(c)	$V_{in}/(1 - \Delta)^2$	3	4	1	4	–
3.16(d)	$V_{in}/(1 - \Delta)^2$	3	5	1	5	–

*NL: number of inductors, NC: number of capacitors, NS: number of switches, ND: number of the diodes, NT: number of windings.

TABLE 3.2
Comparison of Multistage SIBPC Family or HSI-SC Power Converter Family

Figure No.	Switch Voltage Stress	NL	NC	NS	ND	NT
3.18(a)	V_{out}	2	1	1	4	–
3.18(b)	V_{out}	2	2	1	3	–
3.18(c)	$V_{in}/(1 - \Delta)$	2	3	1	3	–
3.18(d)	Max $V_{in}/(1 - \Delta)$	2	3	2	3	–
3.19(a)	$2V_{in}/(1 - \Delta)$	2	4	1	5	–
3.19(b)	$V_{in}(1 + \Delta)/(1 - \Delta)$	3	2	1	4	–
3.19(c)	$2V_{in}/1 - \Delta$	3	3	1	5	–
3.19(d)	$V_{in}(3-\Delta)/(1 - \Delta)$	3	4	2	5	–
3.20(d)	$(V_{out} + V_{in})/2$	2	1	2	1	–
3.20(e)	$V_{out}/2$	2	2	2	2	–
3.20(f)	$(V_{out} - V_{in})/2$	2	3	2	3	–

*NL: number of inductors, NC: number of capacitors, NS: number of switches, ND: number of the diodes, NT: number of windings.

TABLE 3.3

Comparison of Coupled Inductor or Transformer-Based Converter Family

Figure No.	NL	NC	NS	ND	NT
3.21(a)	–	2	1	1	2
3.21(b)	1	2	1	2	2
3.21(c)	–	3	1	3	2
3.21(d)	–	2	1	2	2
3.21(e)	2	4	1	3	2
3.21(f)	2	5	1	5	2
3.22(a)	–	2	2	2	2
3.22(b)	2	4	1	4	2
3.22(c)	1	3	1	4	2
3.22(d)	–	2	1	2	3
3.22(e)	–	3	1	2	3
3.22(f)	1	3	1	2	3
3.23(a)	1	3	1	2	3
3.23(b)	–	1	1	1	2
3.23(c)	–	2	2	1	2

*NL: number of inductors, NC: number of capacitors, NS: number of switches, ND: number of the diodes, NT: number of windings.

TABLE 3.4

Comparison of Luo DC-DC Converter Family

Figure No.	NL	NC	NS	ND	NT
3.25(a)	3	4	2	3	–
3.25(b)	4	5	2	5	–
3.25(c)	5	6	2	7	–
3.25(d)	3	4	1	5	–
3.25(e)	4	5	1	7	–
3.25(f)	5	6	1	9	–
3.26(a)	3	4	1	5	–
3.26(b)	4	5	1	7	–
3.26(c)	5	6	1	9	–

*NL: number of inductors, NC: number of capacitors, NS: number of switches, ND: number of the diodes, NT: number of windings.

TABLE 3.5

Comparison of Z-Source and Quasi Z-Source-Based DC-DC Converter Family

Figure No.	Voltage Gain	NL	NC	NS	ND	NT
3.27	$(1 - \Delta)/(1 - 2\Delta)$	3	3	1	1	–
3.28	$1/(1 - 2\Delta)$	2	3	1	2	–
3.29	$1/(1 - 3\Delta)$	5	6	2	7	–
3.30	$1/(1 - 4\Delta)$	4	7	1	4	–
3.31	$1/(1 - 4\Delta)$	4	7	1	4	–

*NL: number of inductors, NL: number of capacitors, NS: number of switches, ND: number of the diodes, NT: number of windings.

TABLE 3.6

Comparison of Cockcroft Walton Voltage Multiplier-Based Multilevel DC-DC Converter Family

Figure No.	Voltage Stress	NL	NC	NS	ND	NT
3.32(a)	$V_{in}/(1 - \Delta)$	1	$2N + 1$	1	$2N + 1$	–
3.32(b)	$V_{in}/(1 - \Delta)$	1	$2N$	4	$2N$	–
3.32(c)	V_{in}	–	$2N - 2$	2	$2N - 2$	–
3.32(d)	V_{out}/N	1	$2N + 1$	2	$2N + 2$	2
3.33(a)	V_{out}/N	1	$2N - 1$	1	$2N - 1$	–
3.33(b)	$V_{in}/(1 - \Delta)$	1	$2N - 1$	1	$2N - 1$	–
3.23(c)	$V_{in}/(1 - \Delta)$	1	$2N$	1	$2N + 1$	–
3.33(d)	$V_{in}/(1 - \Delta)$	2	$2N$	1	$2N - 1$	–
3.34	$V_{in}(1 + \Delta)/(1 - \Delta)$	2	$2N - 1$	1	$2N + 2$	–

*NL: number of inductors, NC: number of capacitors, NS: number of switches, ND: number of the diodes, NT: number of windings.

3.12 CONCLUSION

Several unidirectional multistage DC-DC converter families were reviewed, based on their boosting type and characteristics of components. DC-DC multistage converters were broadly classified as SCBPC, SIBPC, coupled- and transformer-based converters, Luo DC-DC converter, Z-source converters, and multilevel converter configuration based on the voltage multiplier. Each configuration is explained in detail, and each converter family has its advantages and disadvantages. Based on the review, it can be concluded that the recent converter, again and again, combines

TABLE 3.7

Compared of Converters Configurations in Term of Cost, Efficiency, Reliability, Advantages, Disadvantages and Applications

Converter Configuration	Reliability	Cost	Efficiency	Advantages	Disadvantages
Multistage SCBPC Family	H	A	VH	Small size, low weight, modular	Poor voltage regulation, high start-up current, large number of capacitors
Multistage SIBPC Family	VH	H	H	Very high conversion ratio, merge with many converters	Large number of inductors, weighted, poor performance for high power
Coupled Inductor or Transformer-Based Converter Family	A	VH	H	High power, soft switching, tunable turns ratio, high conversion ratio	Bulky, more losses due to magnetic coupling, complex
Luo DC-DC Converter Family	H	H	A	Very high conversion ratio, average in size	Large number of inductors, capacitors, and diodes; complex in control; high losses
Z-Source Converter Family	H	A	H	Very high conversion ratio, low duty cycle	Not suitable for a higher duty cycle, narrow duty range, large number of components
Cockcroft Walton Voltage Multiplier-Based Multilevel DC-DC Converter Family	A	H	VH	Self-balanced, modular, extendable structure, low number of switches	Large number of capacitors and diodes

A: average, H: high, VH: very high

with the switched capacitor, switched inductor, and numerous boosting methods to advance the performance for DC applications. A comparative study of all converters was presented and, due to low voltage stress across the switch and high boosting capability, HSI-SC, Z-source, and multiplier-based DC-DC converters can be better choices for the photovoltaic applications in terms of cost and efficiency. Disadvantages and advantages of each multistage family were discussed in detail.

4 X-Y Power Converter Family: A New Breed of DC-DC Multistage Configurations for Photovoltaic Applications

4.1 INTRODUCTION

In this chapter, a new buck-boost converter breed named the X-Y power converter family, is described. Recently, several configurations of the X-Y power converter family have been suggested with a voltage double or voltage multiplier for renewable energy applications [212–223]. Initially, in the X-Y power converter family, a total of 16 configurations are suggested, which are highly appropriate for photovoltaic energy applications, and require a DC-DC converter with a high-voltage conversion ratio, such as a photovoltaic-integrated DC-AC multilevel power converters, HVDC applications, and renewable industrial drives [212,213]. Compared to the conventional boost converter and recent DC-DC power converters, the suggested X-Y power converter family can offer a superior high-output voltage with using minimum reactive components and power devices [212]. The universal structure of a X-Y power converter family is discussed. The X-Y power converter family is designed based on the arrangement of reactive elements and semiconductors. The classification of the X-Y converter family is presented in this chapter. The conspicuous features of the X-Y power converter family are discussed in this chapter [179–182]. To investigate the benefits of the X-Y power converter family, all the power converter configurations of the X-Y converter family are compared with recent DC-DC converters. The functionality and feasibility of all the suggested configurations are verified by a numerical simulation software working with Matrix Laboratory 9.0 (R2016a).

4.2 X-Y POWER CONVERTER FAMILY: UNIVERSAL STRUCTURE AND ITS CONVERTER CONFIGURATIONS

The X-Y power converter family (a new breed of the buck-boost converter) is suggested for high-voltage step-up applications to overcome the drawback of cascaded converters and isolated converters [212].

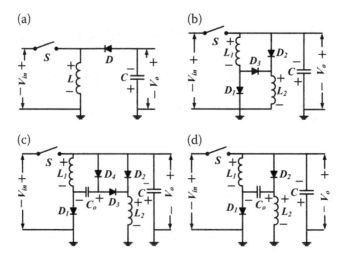

FIGURE 4.1 Power Converter Configurations: (a) Buck-Boost-Converter, (b) Switched-Inductor Buck-Boost Converter, (c) Voltage-Lift Switched-Inductor Converter, (d) Modified Voltage-Lift Switched-Inductor Converter.

For voltage boosting purposes, new buck-boost converters are derived by using an inductor, switched inductor, voltage-lift switched inductor (VLSI), and a modified voltage-lift switched inductor (mVLSI). Figure 4.1 depicts the power converter configurations of (a) buck-boost converter (L converter or BBC), (b) switched-inductor buck-boost converter (2L converter or SI-BBC), (c) voltage-lift switched-inductor converter (2LC converter or VLSI-BBC), and (d) modified voltage-lift switched-inductor converter ($2LC_m$ converter or modified VLSI-BBC) [212–214].

The 2L converter is derived by combining the features of the switched inductor (SI or 2L) and the conventional buck-boost converter (BBC) [215–218]. The 2LC converter is derived by combining the features of the voltage-lift switched inductor (VLSI or 2LC) and conventional BBC [219,220].

The $2LC_m$ converter is derived by combining the features of a modified voltage-lift switched inductor (modified VLSI or $2LC_m$) and conventional BBC [221]. Figure 4.2 depicts the universal structure of the X-Y power converter family [212–223]. Two separate X and Y power converters are used to design the X-Y

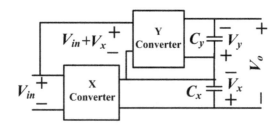

FIGURE 4.2 Universal Structure of X-Y Power Converter Family.

converter family. The X power converter is directly fed by the input voltage source, and the series connection of output voltage of the X power converter and input voltage supply is used to the feed the Y power converter. The inverting sum of the output voltage of Y and X power converters is the total output voltage for the XY converter; (4.1) is used to calculate the output voltage of the X-Y converter,

$$V_o = -\left(V_y + V_x \right) \tag{4.1}$$

where V_y and V_x are the output voltages of Y and X converters, respectively. Based on the number of stages, the whole X-Y converter family is divided into three categories:

- Two-stage X-Y power converter configurations
- Three-stage X-Y power converter configurations
- N-stage L-Y power converter configurations

4.3 TWO-STAGE X-Y POWER CONVERTER CONFIGURATIONS (BASIC X-Y POWER CONVERTER CONFIGURATION)

All of the appropriate combinations of the newly derived BBC are shown in Figure 4.1(a)–(d), and the total 16 power converter configurations are derived in the X-Y power converter family. The details of all the derived two-stage X-Y power converter configurations are shown in Figure 4.3 and Table 4.1; these are the basic configurations of the X-Y power converter family. Two stage X-Y power converter configurations are shown in Figure 4.4(a)–(h) and Figure 4.5(a)–(h).

Further, based on the X converter, all of the 16 basic configurations of the X-Y power converter family are divided into four categories:

- L-Y power converter configuration
- 2L-Y power converter configuration
- 2LC-Y power converter configuration
- 2LC$_m$-Y power converter configuration

The L-Y converter is a category of the X-Y power converter in which the X converter is a conventional BBC or L converter [212–214,222,223]. The 2L-Y converter is a category of the X-Y power converters in which the X converter is a SI-BBC, or 2L, converter or SI-BBC [216–218]. The 2LC-Y converter is a category of the X-Y power converters in which the X converter is a VLSI-BBC, or 2LC, converter [219,220]. The 2LC$_m$-Y converter is a category of the X-Y power converters in which the X converter is a modified VLSI-BBC, or 2LC$_m$, converter [221].

The L-Y power converter configurations of the X-Y family comprise the L-L power converter, L-2L power converter, L-2LC power converter, and L-2LC$_m$ power converter [213]. The L-Y power converter category of the X-Y converter

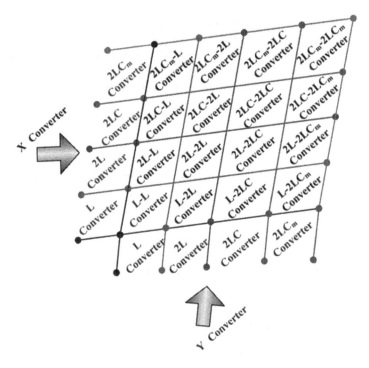

FIGURE 4.3 Two Stage X-Y Power Converter Configurations.

TABLE 4.1

Details of Derived Two Stage X-Y Power Converter Configurations [212]

X-Y Converter		Y DC-DC Converter			
		BBC (L)	**SI (2L)**	**VLSI** **BBC (2LC)**	**Modified VLSI** **BBC (2LC$_m$)**
X DC-DC	BBC (L)	L-L	L-2L	L-2LC	L-2LC$_m$
Converter	SI BBC (2L)	2L-L	2L-2L	2L-2LC	2L-2LC$_m$
	VLSI BBC (2LC)	2LC-L	2LC-2L	2LC-2LC	2LC-2LC$_m$
	Modified VLSI BBC (2LC$_m$)	2LC$_m$-L	2LC$_m$-2L	2LC$_m$-2LC	2LC$_m$-2LC$_m$

family is designed by utilizing two power converters: the L power converter and the Y power converter. The input source directly attached to the L power converter and the series connection of output voltage of the L power converter and the input voltage source is the input for the Y power converter. The inverting sum of the output voltage of the Y and L power converters is the total output voltage for the L-Y converter. (4.2) is used to calculate the output voltage of the L-Y converter. The universal configuration of L-Y converters is shown in Figure 4.6(a).

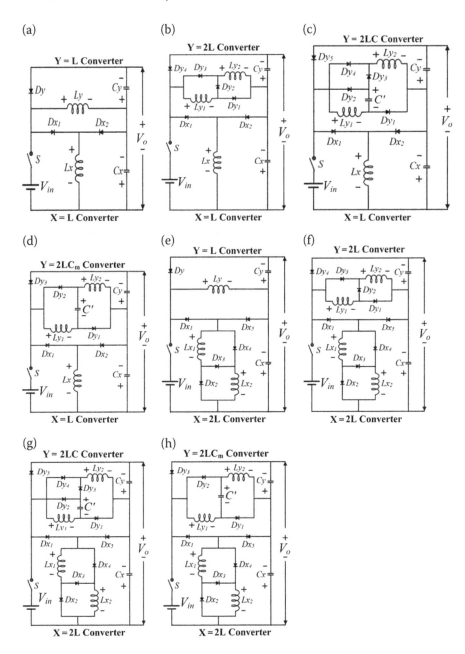

FIGURE 4.4 Power Converter Configuration: (a) L-L Converter, (b) L-2L Converter, (c) L-2LC Converter, (d) L-2LC$_m$ Converter, (e) 2L-L Converter, (f) 2L-2L Converter, (g) 2L-2LC Converter, (h) 2L-2LC$_m$ Converter.

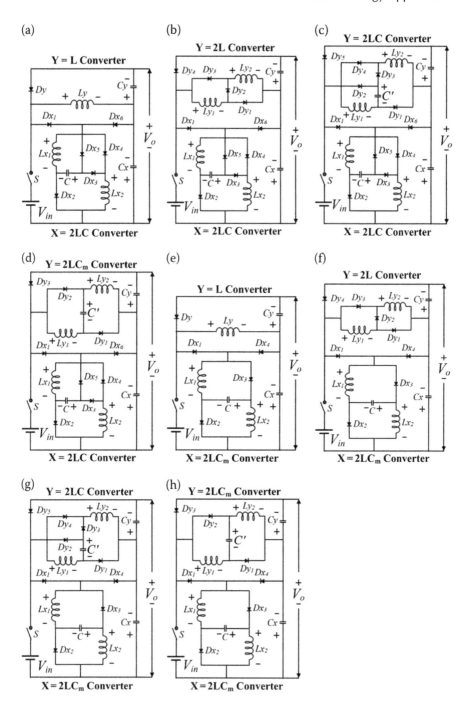

FIGURE 4.5 Power Converter Configuration: (a) 2LC-L Converter, (b) 2LC-2L Converter, (c) 2LC-2LC Converter, (d) 2LC-2LC$_m$ Converter, (e) 2LC$_m$-L Converter, (f) 2LC$_m$-2L Converter, (g) 2LC$_m$-2LC Converter, (h) 2LC$_m$-2LC$_m$ Converter.

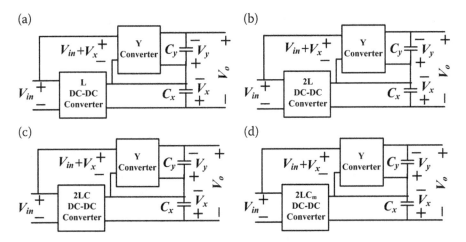

FIGURE 4.6 Power Converter Configuration: (a) Universal Configuration of L-Y Converters, (b) Universal Configuration of 2L-Y Converters, (c) Universal Configuration of 2LC-Y Converters, (d) Universal Configuration of 2LC$_m$-Y Converters.

$$V_o = -\left(V_y + V_L\right)$$ (4.2)

The 2L-Y power converter configurations of the X-Y family comprise the 2L-L power converter, 2L-2L power converter, 2L-2LC power converter, and 2L-2LC$_m$ power converter [215–218]. The 2L-Y power converter category of the X-Y converter family is designed by utilizing two power converters: 2L power converter and Y power converter. The input source directly attached to the 2L power converter and series connection of output voltage of the 2L power converter and the input voltage source is the input for the Y power converter. The inverting sum of the output voltage of the Y and 2L power converters is the total output voltage for the 2L-Y converter. (4.3) is used to calculate the output voltage of the 2L-Y converter. The universal configuration of the 2L-Y converter is shown in Figure 4.6(b).

$$V_o = -\left(V_y + V_{2L}\right)$$ (4.3)

The 2LC-Y power converter configurations of the X-Y family comprise the 2LC-L power converter, 2LC-2L power converter, 2LC-2LC power converter, and 2LC-2LC$_m$ power converter [219,220]. The 2LC-Y power converter category of the X-Y converter family is designed by utilizing two power converters: 2LC power converter and Y power converter. The input source directly attached to the 2LC power converter and series connection of output voltage of the 2LC power converter and the input voltage source is the input for the Y power converter. The inverting sum of the output voltage of Y and 2LC power converters is the total output voltage for the 2LC-Y converter. (4.4) is used to calculate the output voltage of the 2LC-Y converter. The universal configuration of the 2LC-Y converter is shown in Figure 4.6(c).

$$V_o = -\left(V_y + V_{2LC}\right)$$

(4.4)

The 2LC$_m$-Y power converter configurations of the X-Y family comprise the 2LC-L power converter, 2LC$_m$-2L power converter, 2LC$_m$-2LC power converter, and 2LC$_m$-2LC$_m$ power converter [221]. The 2LC$_m$-Y power converter category of the X-Y converter family is designed by utilizing two power converters: 2LC$_m$ power converter and Y power converter. The input source is directly attached to the 2LC$_m$ power converter and series connection of output voltage of the 2LC$_m$ power converter and the input voltage source is the input for the Y power converter. The inverting sum of the output voltage of the Y and 2LC$_m$ power converters is the total output voltage for the 2LC$_m$-Y converter. (4.5) is used to calculate the output voltage of the 2LC$_m$-Y converter. The universal configuration of the 2LC$_m$-Y converter is shown in Figure 4.6(d).

$$V_o = -\left(V_y + V_{2LC_m}\right)$$

(4.5)

4.3.1 WORKING MODES OF THE TWO-STAGE X-Y POWER CONVERTER FAMILY

The two-stage X-Y power converter configuration is divided into two working modes: first when the switch is turned ON (conducting) and another when the switch is turned OFF (not conducting). To illustrate the working mode in the X-Y power converter configurations, the 2LC$_m$-2LC$_m$ converter topology is used. Figure 4.5(h) depicts the power circuit of the 2LC$_m$-2LC$_m$ converter configuration [212].

The 2LC$_m$-2LC$_m$ converter is a combination of two 2LC$_m$ converters. Four capacitors, 4 inductors, and 7 diodes, along with a single switch are needed to design the 2LC$_m$-2LC$_m$ converter. To analyze the converter, steady-state operation is assumed, along with the following:

- Input supply is pure DC
- Component efficiency is 100%; thus, all the components are ideal
- All the inductors of the X converter are the same and identical in rating.
- All the inductors of the Y converter are the same and identical in rating.
- Operating switching frequency is high to reduce voltage ripple of the capacitor.

When switch S is turned ON, input voltage magnetizes the inductors L_{x2} and L_{x1} in parallel via diodes D_{x3} and D_{x2}, respectively. At the same instant, the series connection of voltage across the capacitor, C_x, and input voltage magnetizes the inductors L_{y2} and L_{y1} via diodes D_{y2} and D_{y1}. The C_1 capacitor is charged by an input supply voltage via diodes D_{x3} and D_{x2}.

Similarly, the C' capacitor of the Y converter is charged by the series connection of voltage across capacitor C_x and input voltage. The inverting sum of the output voltage across capacitors C_y and C_x is the total output voltage for the $2LC_m$-$2LC_m$ power converter. Equation (4.6) is used to calculate the output voltage of the $2LC_m$-$2LC_m$ converter. Figure 4.7(a) depicts the equivalent circuit of the $2LC_m$-$2LC_m$ power converter when the switch is turned ON (conducting).

$$\left. \begin{array}{l} V_{Lx_2} = V_{Lx_1} = V_{in} \\ V_{Ly_2} = V_{Ly_1} = V_{in} + V_{Cx} \\ V_C = V_{in} \\ V_{C'} = V_{Cx} + V_{in} \\ V_o = -\left(V_{Cy} + V_{Cx} \right) \end{array} \right\} \tag{4.6}$$

When switch S is turned OFF (not conducting), it detaches an input supply from the main power circuit. Both inductors L_{x2} and L_{x1} are demagnetized in series with capacitor C via load and charges capacitor C_x. Similarly, inductors L_{y1} and L_{y2} are demagnetized in series with capacitor C' to charge capacitor C_y. The inverting sum of the output voltage of voltage across capacitors C_y and C_x is the total output voltage for the $2LC_m$-$2LC_m$ power converter. (4.7) is used to calculate the output voltage of the $2LC_m$-Y converter. Figure 4.7(b) depicts the equivalent circuit of

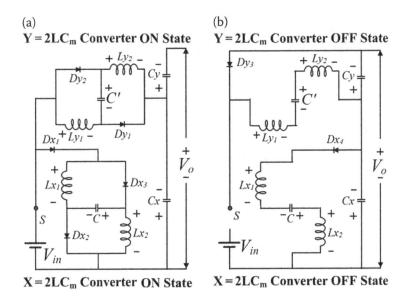

FIGURE 4.7 Power Converter Configurations: (a) Equivalent Circuit of the $2LC_m$-$2LC_m$ Power Converter When Switch Is Turned ON, (b) Equivalent Circuit of the $2LC_m$-$2LC_m$ Power Converter When Switch Is Turned OFF.

the $2LC_m$-$2LC_m$ power converter when the switch S is turned OFF (not conducting).

$$V_o = -\left(V_y + V_{2LC_m}\right) \tag{4.7}$$

From equations (4.6) and (4.7), the voltage conversion ratio of the $2LC_m$-$2LC_m$ power converter is determined and provided in (4.8)

$$\left.\begin{array}{l} V_{Cx} = V_{in}(1 + \Delta)/(1 - \Delta) \\ V_{Cy} = 2V_{in}(1 + \Delta)/(1 - \Delta)^2 \\ V_o = -V_{in}(3 - \Delta)(1 + \Delta)/(1 - \Delta)^2 \end{array}\right\} \tag{4.8}$$

4.3.2 Voltage Conversion Analysis of the Two-Stage X-Y Power Converter Family

In this section, the voltage conversion ratio for all two-stage X-Y power converter configurations is discussed. To analyze the power converter configurations, a steady-state operation assumed, including the following:

- Input supply is pure DC
- All the power devices are practical and V_d is the voltage drop across each device. If $V_d = 0$, then component efficiency is 100% (components are ideal).
- All the inductors of the X converter are the same and identical in rating, and V_d is a voltage drop across each inductor due to internal resistance.
- All the inductors of the Y converter are same and identical in rating, and V_d is a voltage drop across each inductor due to internal resistance.
- Operating switching frequency is high to reduce the voltage ripple of capacitors.

4.3.2.1 Voltage Conversion Ratio of the L-L Power Converter Configuration (X = Y = L)

The power circuit of an L-L power converter configuration is depicted in Figure 4.4(a),

$$\left.\begin{array}{l} V_{Lx} = -3V_d + V_{in} \\ V_{Ly} = -2V_d + (G_{XL} + 1)V_{in} \\ V_o = -(V_{Cy} + V_{Cx}) \end{array}\right\} \text{ON state} \tag{4.9}$$

where G_{XL} is the voltage conversion ratio of the L power converter.

$$\left.\begin{array}{l} V_{Lx} = -2V_d - V_{Cx} \\ V_{Ly} = -2V_d - V_{Cy} \\ V_o = -(V_{Cy} + V_{Cx}) \end{array}\right\} \text{OFF state} \tag{4.10}$$

From, equations (4.9) and (4.10),

$$V_{Cx} = \left(-\frac{(\Delta + 2)V_d}{1 - \Delta} + \frac{V_{in}\Delta}{1 - \Delta} \right) \tag{4.11}$$

$$G_{XL} = \frac{V_{Cx}}{V_{in}} = \left(-\frac{(\Delta + 2)V_d}{(1 - \Delta)V_{in}} + \frac{\Delta}{1 - \Delta} \right) \tag{4.12}$$

$$V_{Cy} = \left(\frac{\Delta(G_{XL} + 1)V_{in}}{1 - \Delta} - \frac{2V_d}{1 - \Delta} \right) \tag{4.13}$$

$$G_{YL} = \frac{V_{Cy}}{V_{in}} = \left(\frac{\Delta(1 + G_{XL})}{1 - \Delta} - \frac{2V_d}{V_{in}(1 - \Delta)} \right) \tag{4.14}$$

where G_{YL} is the voltage conversion ratio of the L power converter, and

$$G_{XY} = G_{LL} = \frac{V_o}{V_{in}} = -(G_{XL} + G_{YL}) \tag{4.15}$$

where G_{XY} is the voltage gain of the L-L power converter.

4.3.2.2 Voltage Conversion Ratio of the L-2L Power Converter Configuration (X = L and Y = 2L)

The power circuit of the L-2L power converter configuration is depicted in Figure 4.4(b),

$$\left.\begin{array}{l} V_{Lx} = -3V_d + V_{in} \\ V_{Ly_1} = -3V_d + (G_{XL} + 1)V_{in} \\ V_{Ly_2} = -3V_d + (G_{XL} + 1)V_{in} \\ V_o = -(V_{Cx} + V_{Cy}) \end{array}\right\} \text{ON state} \tag{4.16}$$

where G_{XL} is the voltage conversion ratio of the L power converter.

$$\left.\begin{array}{l} V_{Lx} = -2V_d - V_{Cx} \\ V_{Ly_2} = V_{Ly_1} = \dfrac{-4V_d - V_{Cy}}{2} \\ V_o = -(V_{Cx} + V_{Cy}) \end{array}\right\} \text{OFF state} \tag{4.17}$$

From equations (4.16) and (4.17),

$$V_{Cx} = \left(\frac{V_{in}\Delta}{1-\Delta} - \frac{V_d(\Delta+2)}{1-\Delta} \right) \tag{4.18}$$

$$G_{XL} = \frac{V_{Cx}}{V_{in}} = \left(\frac{\Delta}{1-\Delta} - \frac{V_d(\Delta+2)}{V_{in}(1-\Delta)} \right) \tag{4.19}$$

$$V_{Cy} = \left(\frac{2V_{in}\Delta(G_{XL}+1)}{1-\Delta} - \frac{2V_d(\Delta+2)}{1-\Delta} \right) \tag{4.20}$$

$$G_{Y2L} = \frac{V_{Cy}}{V_{in}} = \left(\frac{2\Delta(1+G_{XL})}{1-\Delta} - \frac{2(\Delta+2)V_d}{(1-\Delta)V_{in}} \right) \tag{4.21}$$

where G_{Y2L} is the voltage conversion ratio of the 2L power converter and

$$G_{XY} = G_{L-2L} = \frac{V_o}{V_{in}} = -(G_{Y2L} + G_{XL}) \tag{4.22}$$

where G_{XY} is the voltage conversion ratio of the L-2L converter.

4.3.2.3 Voltage Conversion Ratio of the L-2LC Power Converter Configuration (X = L and Y = 2LC)

The power circuit of the L-2LC power converter configuration is depicted in Figure 4.4(c),

$$\left. \begin{array}{l} V_{Lx} = -3V_d + V_{in}, \ V_o = -(V_{Cx} + V_{Cy}) \\ V_{Ly_1} = V_{Ly_2} = V_{C'} = -3V_d + (1+G_{XL})V_{in} \end{array} \right\} \text{ON state} \tag{4.23}$$

where G_{XL} is the voltage conversion ratio of the L power converter.

$$\left. \begin{array}{l} V_{Lx} = -V_{Cx} - 2V_d, \ V_o = -(V_{Cy} + V_{Cx}) \\ V_{Ly_1} = V_{Ly_2} = \frac{V_{C'} - V_{Cy} - 4V_d}{2} \end{array} \right\} \text{OFF state} \tag{4.24}$$

From equations (4.23) and (4.24),

$$V_{Cx} = \left(\frac{V_{in}\Delta}{1-\Delta} - \frac{V_d(\Delta+2)}{1-\Delta} \right) \tag{4.25}$$

$$G_{XL} = \frac{V_{Cx}}{V_{in}} = \left(\frac{\Delta}{1-\Delta} - \frac{V_d(\Delta+2)}{(1-\Delta)V_{in}} \right) \tag{4.26}$$

$$V_{Cy} = \left(\frac{(1 + \Delta)(G_{XL} + 1)V_{in}}{1 - \Delta} + \frac{V_d(\Delta - 7)}{1 - \Delta} \right) \tag{4.27}$$

$$G_{Y2LC} = \frac{V_{Cy}}{V_{in}} = \left(\frac{(G_{XL} + 1)(1 + \Delta)}{1 - \Delta} - \frac{V_d(\Delta - 7)}{V_{in}(1 - \Delta)} \right) \tag{4.28}$$

where G_{Y2LC} is the voltage conversion ratio of the 2LC power converter and

$$G_{XY} = G_{L-2LC} = \frac{V_o}{V_{in}} = -(G_{Y2LC} + G_{XL}) \tag{4.29}$$

where G_{XY} is the voltage conversion ratio of the L-2LC power converter.

4.3.2.4 Voltage Conversion Ratio of the L-2LC$_m$ Power Converter Configuration (X = L and Y = 2LC$_m$)

The power circuit of the L-2LC$_m$ power converter configuration is depicted in Figure 4.4(d),

$$\left. \begin{array}{l} V_{Lx} = -3V_d + V_{in} \\ V_{Ly_1} = V_{Ly_2} = V_{C'} = -3V_d + (1 + G_{XL})V_{in} \\ V_o = -(V_{CY} + V_{CX}) \end{array} \right\} \text{ON state} \tag{4.30}$$

where G_{XL} is the voltage conversion ratio of the L power converter.

$$\left. \begin{array}{l} V_{Lx} = -2V_d - V_{Cx} \\ V_{Ly_1} = V_{Ly_2} = \frac{-V_{Cy} + V_{C'} - 3V_d}{2} \\ V_o = -(V_{Cx} + V_{Cy}) \end{array} \right\} \text{OFF state} \tag{4.31}$$

From equations (4.30) and (4.31),

$$V_{Cx} = \left(\frac{V_{in}\Delta}{1 - \Delta} - \frac{V_d(2 + \Delta)}{1 - \Delta} \right) \tag{4.32}$$

$$G_{XL} = \frac{V_{Cx}}{V_{in}} = \left(\frac{\Delta}{1 - \Delta} - \frac{V_d(2 + \Delta)}{V_{in}(1 - \Delta)} \right) \tag{4.33}$$

$$V_{Cy} = \left(\frac{(1 + G_{XL})(1 + \Delta)V_{in}}{1 - \Delta} - \frac{6V_d}{1 - \Delta} \right) \tag{4.34}$$

$$G_{Y2LC_m} = \frac{V_{Cy}}{V_{in}} = \left(\frac{(1 + G_{XL})(1 + \Delta)}{1 - \Delta} - \frac{6V_d}{V_{in}(1 - \Delta)} \right) \tag{4.35}$$

where is the voltage conversion ratio of the 2LC$_m$ converter and

$$G_{XY} = G_{L-2LC_m} = \frac{V_o}{V_{in}} = -(G_{Y2LC_m} + G_{XL}) \qquad (4.36)$$

where G_{XY} is the voltage conversion ratio of the L-2LC$_m$ converter.

Similarly, the previous method also applies to find the voltage conversion ratio of the two-stage X-Y power converter family is derived while considering the voltage drop across semiconductor devices and the voltage drop across an internal resistance of the inductor. In Table 4.2, the voltage conversion ratio of all two-stage X-Y converter configurations is provided, along with the voltage conversion ratio of the recently

---

TABLE 4.2

Voltage Conversion Ratios of Two-Stage X-Y Power Converter Configurations and Recently Discussed Converters

Converter Topology	Conversion Ratio
L-L Converter	$(\Delta^2-2\Delta)\,/(1-\Delta)^2$
L-2L Converter	$(\Delta^2-3\Delta)/(1-\Delta)^2$
L-2LC Converter	$(\Delta^2-2\Delta-1)/(1-\Delta)^2$
L-2LC$_m$ Converter	$(\Delta^2-2\Delta-1)/(1-\Delta)^2$
2L-L Converter	$(\Delta^2-3\Delta)/(1-\Delta)^2$
2L-2L Converter	$-4\Delta/(1-\Delta)^2$
2L-2LC Converter	$(\Delta^2-4\Delta-1)/(1-\Delta)^2$
2L-2LC$_m$ Converter	$(\Delta^2-4\Delta-1)/(1-\Delta)^2$
2LC-L Converter	$(\Delta^2-2\Delta-1)/(1-\Delta)^2$
2LC-2L Converter	$(\Delta^2-4\Delta-1)/(1-\Delta)^2$
2LC-2LC Converter	$(\Delta^2-2\Delta-3)/(1-\Delta)^2$
2LC-2LC$_m$ Converter	$(\Delta^2-2\Delta-3)/(1-\Delta)^2$
2LC$_m$-L Converter	$(\Delta^2-2\Delta-1)/(1-\Delta)^2$
2LC$_m$-2L Converter	$(\Delta^2-4\Delta-1)/(1-\Delta)^2$
2LC$_m$-2LC Converter	$(\Delta^2-2\Delta-3)/(1-\Delta)^2$
2LC$_m$-2LC$_m$ Converter	$(\Delta^2-2\Delta-3)/(1-\Delta)^2$
Conventional Boost Converter	$1/(1-\Delta)$
SI Boost Converter	$1 + \Delta/(1-\Delta)$
Single-Switch Quadratic Boost Converter	$1/(1-\Delta)^2$
Conventional Three-Level Boost Converter	$2/(1-\Delta)$
Quadratic Three-Level Boost Converter	$1/(1-\Delta)^2$
Converters Using Bootstrap Capacitors and Boost Inductors	$3 + \Delta/1-\Delta$
SI-BBC	$1 + \Delta/1-\Delta$
Two-Phase Quadrupled Interleaved Boost Converter	$4/(1 - \Delta)$
High-Voltage Gain Two-Phase Interleaved Boost Converter Using One VMC	$((VMC + 1)/1 - \Delta)$
Extra High Voltage (HV) DC-DC Converter	$4/(1-\Delta)$

discussed converters given for comparison. We see that the slope of the inductor current is positive when the switch is turned ON and negative when the switch is turned OFF.

4.3.3   Validation of the Two-Stage X-Y Power Converter Configurations

The functionality and concept of all 16 configurations of the two-stage X-Y power converter family are verified by the simulation software Matrix Laboratory 9.0 (R2016a), when considering an input supply of 10 V, power of 100 W, a duty cycle of 60%, and a switching frequency of 50 kHz. The waveform of output voltage for all the two-stage X-Y converter configurations is depicted in Figures 4.8(a)–(h) and 4.9(a)–(h). We find that every configuration of a two-stage XY converter provides a negative output voltage.

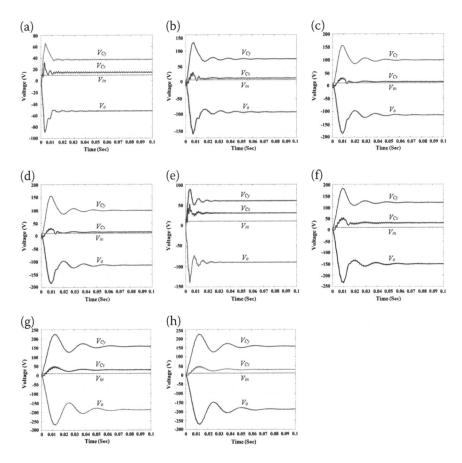

FIGURE 4.8 Simulation Results of Input Voltage, Output Voltage, and Voltage Across Capacitor: (a) L-L, (b) L-2L, (c) L-2LC, (d) L-2LC$_m$, (e) 2L-L, (f) 2L-2L, (g) 2L-2LC, (h) 2L-2LC$_m$.

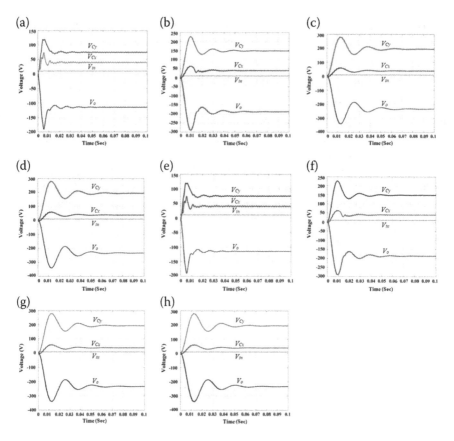

FIGURE 4.9 Simulation Results of Input Voltage, Output Voltage, and Voltage Across Capacitor: (a) 2LC-L, (b) 2LC-2L, (c) 2LC-2LC, (d) 2LC-2LC$_m$, (e) 2LC$_m$-L, (f) 2LC$_m$-2L, (g) 2LC$_m$-2LC, (h) 2LC$_m$-2LC$_m$.

The voltage conversion ratio of all the configurations of X-Y power converters is higher than the conversion ratio of the conventional boost converter and recent boost converter (Table 4.2). The 2LC-2LC, 2LC-2LC$_m$, 2LC$_m$-2LC, and 2LC$_m$-2LC$_m$ power converter configurations provide a maximum voltage conversion ratio in the two-stage X-Y power converter family. The 2LC$_m$-2LC$_m$ power converter achieves a voltage conversion ratio of 24 at a 60% duty cycle.

4.4 THREE-STAGE X-Y POWER CONVERTER CONFIGURATIONS (BASIC X-Y POWER CONVERTER CONFIGURATION WITH VOLTAGE DOUBLER)

To increase the voltage conversion ratio, a voltage doubler is attached to the existing X-Y power converter family, which forms a three-stage X-Y power converter family. Sixteen new topologies are derived by using a combination of a voltage doubler and the X-Y power converter family. Three-stage X-Y power converter configurations are shown in Figures 4.10(a)–(h) and 4.11(a)–(h). Further, based on

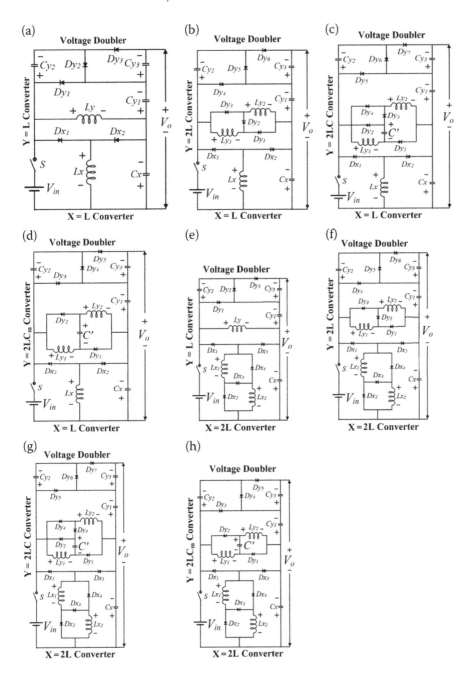

FIGURE 4.10 Power Converter Configurations: (a) L-LVD, (b) L-2LVD, (c) L-2LCVD, (d) L-2LC$_m$ VD, (e) 2L-LVD, (f) 2L-2LVD, (g) 2L-2LCVD, (h) 2L-2LC$_m$VD.

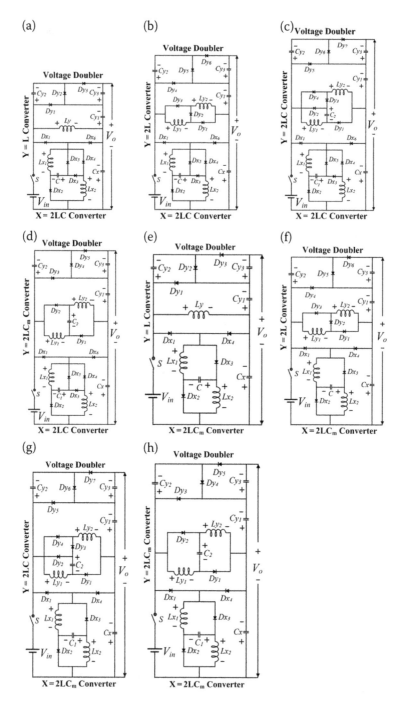

FIGURE 4.11 Power Converter Configuration: (a) 2LC-LVD, (b) 2LC-2LVD, (c) 2LC-2LCVD, (d) 2LC-2LC$_m$VD, (e) 2LC$_m$-LVD, (f) 2LC$_m$-2LVD, (g) 2LC$_m$-2LCVD, (h) 2LC$_m$-2LC$_m$VD.

the X converter, all the three-stage X-Y power converter configurations are divided into four categories [213,216,218–221]:

- L-Y power converter configuration with voltage doubler (three-stage L-Y power converter)
- 2L-Y power converter configuration with voltage doubler (three-stage 2L-Y power converter)
- 2LC-Y power converter configuration with voltage doubler (three-stage 2LC-Y power converter)
- 2LC$_m$-Y power converter configuration with voltage doubler (three-stage 2LC$_m$-Y power converter)

Three-stage L-Y power converter configurations include the L-LVD power converter, L-2LVD power converter, L-2LCVD power converter, and L-2LC$_m$VD power converter [213,223]. In L-LVD, LVD is considered a Y converter (two-stage converter), which combines the features of the L converter and voltage doubler. In L-2LVD, 2LVD is considered a two-stage Y converter, which combines the features of the 2L converter and voltage doubler. In L-2LCVD, 2LCVD is considered a two-stage Y converter, which combines the features of the 2LC converter and voltage doubler. In L-2LC$_m$VD, 2LC$_m$VD is considered a two-stage Y converter, which combines the features of the 2LC$_m$ converter and voltage doubler. This three-stage L-Y power converter category of the X-Y converter family is designed by utilizing two power converters with a voltage doubler: L power converter and voltage-doubler-based Y power converter (two-stage Y converter).

The input source is directly attached to the L power converter and the series connection of output voltage of the L power converter and input voltage source is the input for the two-stage Y power converter. The inverting sum of the output voltage of the two-stage Y converter and L power converter is the total output voltage for the three-stage L-Y converter.

Three-stage 2L-Y power converter configurations include the 2L-LVD power converter, 2L-2LVD power converter, 2L-2LCVD power converter, and 2L-2LC$_m$VD power converter [216,218]. In 2L-LVD, LVD is considered a Y converter (two-stage converter), which combines the features of the L converter and voltage doubler. In 2LCm-2LVD, 2LVD is considered a two-stage Y converter, which combines the features of the 2L converter and voltage doubler. In 2L-2LCVD, 2LCVD is considered a two-stage Y converter, which combines the features of the 2LC converter and voltage doubler. In 2L-2LC$_m$VD, 2LC$_m$VD is considered a two-stage Y converter, which combines the features of the 2LC$_m$ converter and voltage doubler. This three-stage 2L-Y power converter category of the X-Y converter family is designed by utilizing two power converters with a voltage doubler: 2L power converter and voltage-doubler-based Y power converter (two-stage Y converter). The input source is directly attached to the 2L power converter and series connection of the output voltage of the 2L power converter and the input voltage source is the input for the two-stage Y power converter. The inverting sum of the output voltage of the two-stage Y converter and 2L power converter is the total output voltage for the three-stage 2L-Y converter.

Three-stage 2LC-Y power converter configurations include the 2LC-LVD power converter, 2LC-2LVD power converter, 2LC-2LCVD power converter, and 2LC-2LC$_m$VD power converter [219,220]. In 2LC-LVD, LVD is considered a Y converter (two-stage converter), which combines the features of the L converter and voltage doubler. In 2LC-2LVD, 2LVD is considered a two-stage Y converter, which combines the features of the 2L converter and voltage doubler. In 2LC-2LCVD, 2LCVD is considered a two-stage Y converter, which combines the features of the 2LC converter and voltage doubler. In 2LC-2LC$_m$VD, 2LC$_m$VD is considered a two-stage Y converter, which combines the features of the 2LC$_m$ converter and voltage doubler. This three-stage 2LC-Y power converter category of the X-Y converter family is designed by utilizing two power converters with a voltage doubler: 2LC power converter and voltage-doubler-based Y power converter (two-stage Y converter). The input source directly attaches to the 2LC power converter, and the series connection of output voltage of the 2LC power converter and the input voltage source is the input for two-stage Y power converter. The inverting sum of the output voltage of the two-stage Y converter and 2LC power converter is the total output voltage for the three-stage 2LC-Y converter.

Three-stage 2LC$_m$-Y power converter configurations include the 2LC$_m$-LVD power converter, 2LC$_m$-2LVD power converter, 2LC$_m$-2LCVD power converter, and 2LC$_m$-2LC$_m$VD power converter [221]. In 2LC$_m$-LVD, LVD is considered a Y converter (two-stage converter), which combines the features of the L converter and voltage doubler. In 2LC$_m$-2LVD, 2LVD is considered a two-stage Y converter, which combines the features of the 2L converter and voltage doubler. In 2LC$_m$-2LCVD, 2LCVD is considered a two-stage Y converter, which combines the features of the 2LC converter and voltage doubler. In 2LC$_m$-2LC$_m$VD, 2LC$_m$VD is considered a two-stage Y converter, which combines the features of the 2LC$_m$ converter and voltage doubler. This three-stage 2LC$_m$-Y power converter category of the X-Y converter family is designed by utilizing two power converters with a voltage doubler: 2LC$_m$ power converter and voltage-doubler-based Y power converter (two-stage Y converter). The input source is directly attached to the 2LC$_m$ power converter, and the series connection of output voltage of the 2LC$_m$ power converter and input voltage source is the input for two-stage Y power converter. The inverting sum of the output voltage of the two-stage Y converter and 2LC$_m$ power converter is the total output voltage for the three-stage 2LC$_m$-Y converter.

4.4.1 WORKING MODES OF THE THREE-STAGE X-Y POWER CONVERTER FAMILY

The three-stage X-Y power converter configuration is divided into two modes: first, when the switch turned ON (conducting), and another when switch is turned OFF (not conducting). To explain the working mode of the three-stage X-Y power converter configurations, the 2LC$_m$-2LC$_m$VD converter configuration is considered.

Figure 4.11(h) depicts the power circuit of the 2LC$_m$-2LC$_m$VD converter configuration. The 2LC$_m$-2LC$_m$VD converter is a combination of two 2LC$_m$ converters and a voltage doubler. A total of 6 capacitors, 4 inductors, and 9 diodes along with a single switch are needed to design the $2LC_m$-$2LC_mVD$ converter configuration. To analyze the converter, a steady-state operation is assumed along with the following:

- Input supply is pure DC
- Component efficiency is 100%; thus, all the components are ideal
- All the inductors of the X converter are the same and identical in rating.
- All the inductors of the Y converter are the same and identical in rating.
- Operating switching frequency is high to reduce the voltage ripple of the capacitor.

When switch S is in an ON state (conducting), the L_{x1} inductor is magnetized by input supply (V_{in}) via uncontrolled devices D_{x1} and D_{x2} and a controlled device S, whereas the L_{x2} inductor is magnetized by input supply (V_{in}) via uncontrolled devices D_{x1} and D_{x3} and a controlled device S. Via uncontrolled devices D_{x1}, D_{x2}, and D_{x3} and a controlled device S, capacitor C_1 is charged by the input voltage (V_{in}). The L_{y1} inductor is magnetized by the capacitor C_x voltage and input supply (V_{in}) via uncontrolled device D_{y1} and a controlled device S, whereas inductor L_{y2} is charged by voltage across capacitor C_x and input supply (V_{in}) via uncontrolled device D_{y2} and a controlled device S. Via uncontrolled devices D_{y1} and D_{y2} and a controlled device S, capacitor C_2 is charged by voltage capacitor C_x and input voltage (V_{in}).

Via uncontrolled device D_{y4} and a controlled device S, capacitor C_{y2} is charged by the input voltage (V_{in}) and voltage of capacitors C_{y1} and C_x. Therefore, by inverting the addition of capacitor C_x and C_y ($C_y = C_{y1} + C_{y3}$), voltage is the output voltage of $2LC_m$-$2LC_mVD$ power converter. Figure 4.12(a) depicts the equivalent circuit of the $2LC_m$-$2LC_mVD$ power converter when the switch is turned ON (conducting).

The main power supply is disconnected from the power circuit of the $2LC_m$ converter when switch S is in the OFF state (not conducting). L_{x1} and L_{x2} inductors are demagnetized in series with capacitor C_1 via uncontrolled device D_{x4} to transfer its energy to capacitor C_x. At the same time, L_{y1} and L_{y2} inductors are demagnetized in series with capacitor C_2 via uncontrolled device D_{Y3} to transfer its energy to charge the capacitor C_{y1}. The C_{y2} capacitor transferred its energy to charge the C_{y3} capacitor via uncontrolled device D_{y5}. Therefore, inverting the addition of capacitor C_x and C_y ($C_y = C_{y1} + C_{y3}$) voltages is the output voltage of $2LC_m$-$2LC_mVD$ power converter. We observe that all the inductors available in the suggested $2LC_m$-$2LC_mVD$ converter are magnetized when the control switch is in the ON state and demagnetized when the control switch is in the OFF state. Figure 4.12(b) depicts the equivalent circuit of the $2LC_m$-$2LC_mVD$ power converter when the switch is turned OFF (not conducting).

4.4.2 Voltage Conversion Analysis of the Three-Stage X-Y Power Converter Family

In this section, the voltage conversion ratio for all three-stage X-Y power converter configurations are discussed. To analyze the power converter configurations, the steady-state operation is assumed, along with the following:

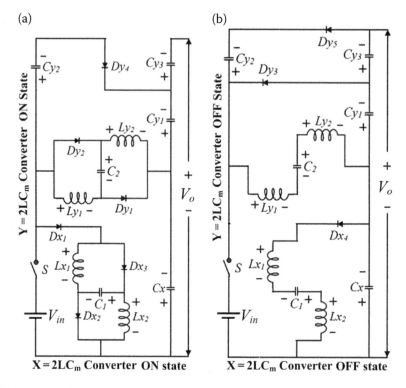

FIGURE 4.12 Power Converter Configurations: (a) Equivalent Circuit of the $2LC_m$-$2LC_m VD$ Power Converter When a Switch Is Turned ON, (b) Equivalent Circuit of the $2LC_m$-$2LC_m VD$ Power Converter When a Switch Is Turned OFF.

- Input supply is pure DC
- All the power devices are practical and V_d is the voltage drop across each device. If $V_d = 0$ then component efficiency is 100% (all the components are ideal).
- All the inductors of the X converter are the same and identical in rating, and V_d is a voltage drop across each inductor because of internal resistance.
- All the inductors of the Y converter are the same and identical in rating, and V_d is a voltage drop across each inductor for internal resistance.
- Operating switching frequency is high to reduce the voltage ripple of the capacitor.

4.4.2.1 Voltage Conversion Ratio of the L-LVD Power Converter Configuration

The power circuit of the L-LVD power converter configuration is depicted in Figure 4.10(a),

$$V_{Cx} = \left(\frac{V_{in}\Delta}{1 - \Delta} - \frac{V_d(\Delta + 2)}{1 - \Delta} \right) \tag{4.37}$$

$$G_X = G_{XL} = \frac{V_{Cx}}{V_{in}} = \left(\frac{\Delta}{1 - \Delta} - \frac{V_d(\Delta + 2)}{(1 - \Delta)V_{in}} \right)$$ (4.38)

where G_{XL} is the voltage conversion ratio of the L converter, and

$$V_{Cy_1} = \left(\frac{(G_{XL} + 1)\Delta V_{in}}{1 - \Delta} - \frac{2V_d}{1 - \Delta} \right)$$ (4.39)

$$G_{Y1} = \frac{V_{Cy_1}}{V_{in}} = \left(\frac{(G_{XL} + 1)\Delta}{1 - \Delta} - \frac{2V_d}{(1 - \Delta)V_{in}} \right)$$ (4.40)

where G_{Y1} is the ratio of voltage across C_1 and the input supply voltage, and

$$G_{Y2} = \frac{V_{Cy_2}}{V_{in}} = \left(1 + G_{XL} + G_{Y1} - \frac{4V_d}{V_{in}} \right)$$ (4.41)

where G_{Y2} is the ratio of voltage across C_2 and the input supply voltage, and

$$G_Y = G_{LVD} = G_{Y2} + G_{Y1}$$ (4.42)

$$G_{XY} = G_{L-LVD} = \frac{V_o}{V_{in}} = -(G_X + G_Y)$$ (4.43)

where G_{XY} is the voltage conversion ratio of the L-LVD converter.

4.4.2.2 Voltage Conversion Ratio of the L-2LVD Power Converter Configuration

The power circuit of the L-2LVD power converter configuration is depicted in Figure 4.10(b),

$$V_{CX} = \left(\frac{V_{in}\Delta}{1 - \Delta} - \frac{V_d(\Delta + 2)}{1 - \Delta} \right)$$ (4.44)

$$G_{XL} = \frac{V_{CX}}{V_{in}} = \left(\frac{\Delta}{1 - \Delta} - \frac{V_d(\Delta + 2)}{(1 - \Delta)V_{in}} \right)$$ (4.45)

where G_{XL} is the voltage conversion ratio of the L converter, and

$$V_{Cx} = \left(\frac{V_{in}\Delta}{1 - \Delta} - \frac{V_d(\Delta + 2)}{1 - \Delta} \right)$$ (4.46)

$$G_{XL} = \frac{V_{Cx}}{V_{in}} = \left(\frac{\Delta}{1 - \Delta} - \frac{V_d(\Delta + 2)}{(1 - \Delta)V_{in}} \right) \tag{4.47}$$

$$V_{Cy_1} = \left(\frac{2\Delta(G_{XL} + 1)V_{in}}{1 - \Delta} - \frac{2V_d(\Delta + 2)}{1 - \Delta} \right) \tag{4.48}$$

$$G_{Y1} = \frac{V_{Cy_1}}{V_{in}} = \left(\frac{2(G_{XL} + 1)\Delta}{1 - \Delta} - \frac{2V_d(\Delta + 2)}{(1 - \Delta)V_{in}} \right) \tag{4.49}$$

$$G_{Y2} = \frac{V_{Cy_2}}{V_{in}} = \left(1 + G_{Y1} + G_{XL} - \frac{6V_d}{V_{in}} \right) \tag{4.50}$$

where G_{Y2} is the ratio of voltage across C_2 and the input supply voltage, and

$$G_Y = G_{2LVD} = G_{Y1} + G_{Y2} \tag{4.51}$$

$$G_{XY} = G_{L-2LVD} = \frac{Vo}{V_{in}} = -(G_X + G_Y) \tag{4.52}$$

where G_{XY} is the voltage conversion ratio of the L-2LVD converter.

4.4.2.3 Voltage Conversion Ratio of the L-2LCVD Power Converter Configuration

The power circuit of the L-2LCVD power converter configuration is depicted in Figure 4.10(c),

$$V_{Cx} = \left(\frac{V_{in}\Delta}{1 - \Delta} - \frac{V_d(\Delta + 2)}{1 - \Delta} \right) \tag{4.53}$$

$$G_{XL} = \frac{V_{Cx}}{V_{in}} = \left(\frac{\Delta}{1 - \Delta} - \frac{V_d(\Delta + 2)}{(1 - \Delta)V_{in}} \right) \tag{4.54}$$

where G_{XL} is the voltage conversion ratio of the L power converter, and

$$V_{Cx} = \left(\frac{V_{in}\Delta}{1 - \Delta} - \frac{V_d(\Delta + 2)}{1 - \Delta} \right) \tag{4.55}$$

$$G_{XL} = \frac{V_{Cx}}{V_{in}} = \left(\frac{\Delta}{1 - \Delta} - \frac{V_d(2 + \Delta)}{(1 - \Delta)V_{in}} \right) \tag{4.56}$$

$$V_{Cy_1} = \left(\frac{(G_{XL} + 1)(1 + \Delta)V_{in}}{1 - \Delta} + \frac{(\Delta - 7)V_d}{1 - \Delta} \right) \tag{4.57}$$

$$G_{Y1} = \frac{V_{Cy_1}}{V_{in}} = \left(\frac{(1 + \Delta)(G_{XL} + 1)}{1 - \Delta} - \frac{V_d(\Delta - 7)}{(1 - \Delta)V_{in}} \right) \tag{4.58}$$

where G_{Y1} is the ratio of voltage across C_1 and the input supply voltage, and

$$G_{Y2} = \frac{V_{Cy_2}}{V_{in}} = \left(1 + G_{Y1} + G_{XL} - \frac{6V_d}{V_{in}} \right) \tag{4.59}$$

where G_{Y2} is the ratio of voltage across C_2 and the input supply voltage, and

$$G_Y = G_{2LCVD} = G_{Y1} + G_{Y2} \tag{4.60}$$

$$G_{XY} = G_{L-2LCVD} = \frac{V_o}{V_{in}} = -(G_X + G_Y) \tag{4.61}$$

where G_{XY} is the voltage conversion ratio of the L-2LCVD converter.

4.4.2.4 Voltage Conversion Ratio of the L-2LC$_m$VD Power Converter Configuration

The power circuit of the L-2LC$_m$VD power converter configuration is depicted in Figure 4.10(d),

$$V_{CX} = \left(\frac{V_{in}\Delta}{1 - \Delta} - \frac{V_d(\Delta + 2)}{1 - \Delta} \right) \tag{4.62}$$

$$G_{XL} = \frac{V_{Cx}}{V_{in}} = \left(\frac{\Delta}{1 - \Delta} - \frac{V_d(\Delta + 2)}{(1 - \Delta)V_{in}} \right) \tag{4.63}$$

$$V_{Cy_1} = \left(\frac{(1 + \Delta)(1 + G_{XL})V_{in}}{1 - \Delta} - \frac{6V_d}{1 - \Delta} \right) \tag{4.64}$$

$$G_{Y1} = \frac{V_{Cy_1}}{V_{in}} = \left(\frac{(1 + \Delta)(1 + G_{XL})}{1 - \Delta} - \frac{6V_d}{(1 - \Delta)V_{in}} \right) \tag{4.65}$$

where G_{Y1} is the ratio of voltage across C_1 and the input supply voltage, and

$$G_{Y2} = \frac{V_{Cy_2}}{V_{in}} = \left(1 + G_{XL} + G_{Y1} - \frac{6V_d}{V_{in}} \right) \tag{4.66}$$

where G_{Y2} is the ratio of voltage across C_2 and the input supply voltage, and

$$G_Y = G_{2LC_mVD} = G_{Y1} + G_{Y2} \qquad (4.67)$$

$$G_{XY} = G_{L-2LC_mVD} = \frac{V_o}{V_{in}} = -(G_X + G_Y) \qquad (4.68)$$

where G_{XY} is the voltage conversion ratio of the L-2LC$_m$VD converter. Similarly, the voltage conversion ratio for all three-stage X-Y power converter configurations is derived. It is observed that L-2LCVD and L-2LC$_m$VD provide the maximum voltage conversion ratio in the three-stage L-Y configuration category. In addition, 2L-2LCVD and 2L-2LC$_m$VD provide the maximum voltage conversion ratio in the three-stage 2L-Y configuration category. Further, 2LC-2LCVD and 2LC-2LC$_m$VD provide the maximum voltage conversion ratio in the three-stage 2LC-Y configuration category. Furthermore, 2LC$_m$-2LCVD and 2LC$_m$-2LC$_m$VD provide the maximum voltage conversion ratio in the three-stage 2LC$_m$-Y configuration category.

4.4.3 VALIDATION OF THE THREE-STAGE X-Y POWER CONVERTER CONFIGURATIONS

The functionality and the concept of all 16 configurations of the three-stage X-Y power converter family are verified by simulation software Matrix Laboratory 9.0 (R2016a), with an input supply of 10 V, power of 100 W, duty cycle at 60%, and switching frequency at 50 kHz.

The waveform of the output voltage and input voltage for all three-stage L-Y power converter configurations is depicted in Figure 4.13(a)–(d). All three-stage L-Y power converter configurations have inverting output voltage. The L-LVD converter has a larger voltage conversion ratio compared to the L-L converter. The L-2LVD converter has a larger voltage conversion ratio compared to the L-2L converter. The L-2LCVD converter has a larger voltage conversion ratio compared to the L-2LC converter and the L-2LC$_m$VD converter has a larger voltage conversion ratio compared to the L-2LC$_m$ power converter. Three-stage L-Y power converter configurations have larger voltage conversion ratios compared to two-stage L-Y converters. It is also observed that L-2LCVCD and L-2LC$_m$VCD converter configurations have a maximum conversion ratio in three-stage L-Y power converter configurations.

The waveform of the output voltage and input voltage for all three-stage 2L-Y power converter configurations is depicted in Figure 4.13(e)–(h) (here, controlled input is considered for simulation). All three-stage 2L-Y power converter configurations have inverting output voltage. In three-stage 2L-Y configurations, the 2L-2LC$_m$VD power converter provides the higher negative voltage conversion ratio (−39). The voltage conversion ratio of 2L-2LVD and L-2LVD is −31 and −19, respectively. The output voltage of 2L-2LC$_m$VD, 2L-2LCVD, 2L-2LVD, and 2L-LVD converters is −388, −388, −309, and −189 V, respectively.

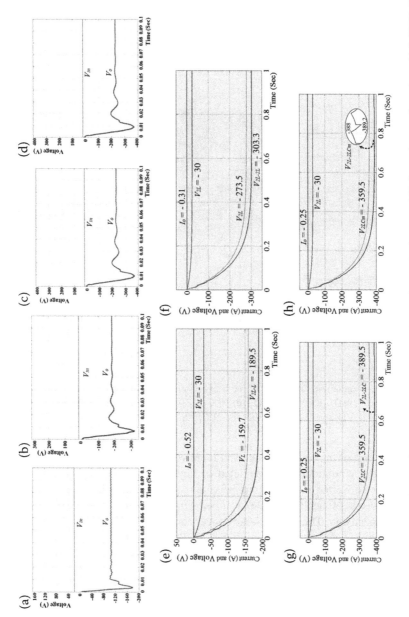

FIGURE 4.13 Simulation Results of Three-Stage Converters, Input Voltage, Output Voltage, and Voltage Across Capacitor: (a) L-LVD (b) L-2LVD, (c) L-2LCVD, (d) L-2LC$_m$VD, (e) 2L-LVD, (f) 2L-LVD, (g) 2L-2LCVD, (h) 2L-2LC$_m$VD.

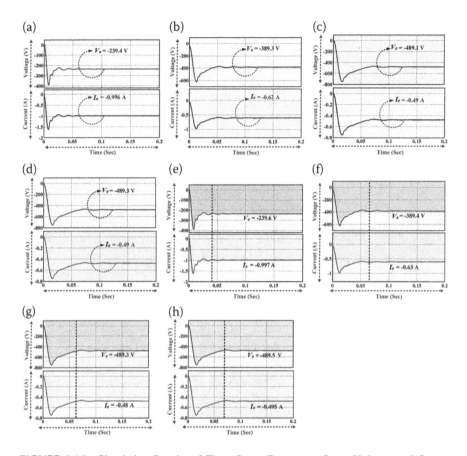

FIGURE 4.14 Simulation Results of Three-Stage Converters, Input Voltage, and Output Voltage: (a) 2LC-LVD, (b) 2LC-2LVD, (c) 2LC-2LCVD, (d) 2LC-2LC$_m$VD, (e) 2LC$_m$-LVD, (f) 2LC$_m$-2LVD, (g) 2LC$_m$-2LCVD, (h) 2LC$_m$-2LC$_m$VD.

The waveform of the output voltage and input voltage for all three-stage 2LC-Y power converter configurations is depicted in Figure 4.14(a)–(d). All three-stage 2LC-Y power converter configurations have inverting output voltage. The output voltage and current of the 2LC-LVD power converter configuration is –239.4 V and –0.996 A, respectively. Thus, the voltage conversion ratio is –24; achieved by using a 2LC-LVD power converter configuration at a 60% duty cycle.

The output voltage and current of 2LC-2LVD is –389.3 V and –0.62 A, respectively. Thus, the voltage conversion ratio of –39 is achieved by using the 2LC-2LVD power converter configuration at a 60% duty cycle. The output voltage and current of 2LC-2LCVD is –489.1 V and –0.49 A, respectively. Thus, the voltage conversion ratio of –49 is achieved by using a 2LC-2LCVD power converter configuration at a 60% duty cycle. The output voltage and current of 2LC-2LC$_m$VD is –489.1 V and –0.49 A, respectively. Thus, the voltage conversion ratio of –49 is achieved by using a 2LC-2LC$_m$VD power converter configuration at a 60% duty cycle. The 2LC-LVD converter has a larger voltage conversion ratio compared to the 2LC-L converter. The 2LC-2LVD converter has a larger voltage conversion ratio compared to the 2LC-2L converter. The 2LC-2LCVD converter has a larger voltage conversion ratio compared to the 2LC-2LC converter, and the 2LC-2LC$_m$VD converter has a larger voltage conversion ratio compared to the 2LC-2LC$_m$ power converter. Three-stage 2LC-Y power converter configurations have a larger voltage conversion ratio compared to a two-stage 2LC-Y converter. The 2LC-2LCVCD and 2LC-2LC$_m$VCD converter configurations have a maximum conversion ratio in three-stage 2LC-Y power converter configurations.

The waveform of the output voltage and input voltage for all three-stage 2LC$_m$-Y power converter configurations is depicted in Figure 4.14(e)–(h). All three-stage 2LC$_m$-Y power converter configurations have inverting output voltage. The output voltage and current of the 2LC$_m$-LVD power converter configuration is –239.6 V and –0.997 A, respectively. Thus, the voltage conversion ratio of –24 is achieved by using the 2LC$_m$-LVD power converter configuration at a 60% duty cycle. The output voltage and current of the 2LC$_m$-2LVD is –389.4 V and –0.63 A, respectively.

Thus, the voltage conversion ratio of –39 is achieved by using the 2LC$_m$-2LVD power converter configuration at a 60% duty cycle. The output voltage and current of 2LC$_m$-2LCVD is –489.3 V and –0.48 A, respectively. Thus, the voltage conversion ratio of –49 is achieved by using a 2LC$_m$-2LCVD power converter configuration at a 60% duty cycle. The output voltage and current of a 2LC$_m$-2LC$_m$VD is –489.5 V and –0.495 A, respectively. Thus, the voltage conversion ratio of –49 is achieved by using a 2LC$_m$-2LC$_m$VD power converter configuration at a 60% duty cycle. A 2LC$_m$-LVD converter has a larger voltage conversion ratio compared to the 2LC$_m$-L converter, a 2LC$_m$-2LVD converter has a larger voltage conversion ratio compared to the 2LC$_m$-2L converter, a 2LC$_m$-2LCVD converter has a larger voltage conversion ratio compared to the 2LC$_m$-2LC converter, and a 2LC$_m$-2LC$_m$VD converter has a larger voltage conversion ratio compared to the 2LC$_m$-2LC$_m$ power converter. Three-stage 2LC$_m$-Y power converter configurations have a larger voltage conversion ratio compared to the two-stage 2LC$_m$-Y converter. 2LC$_m$-2LCVCD and 2LC$_m$-2LC$_m$VCD converter configurations have a maximum conversion ratio in three-stage 2LC$_m$-Y power converter configurations.

4.5 N-STAGE L-Y POWER CONVERTER CONFIGURATIONS (L-Y MULTILEVEL BOOST CONVERTER, L-Y MBC)

In this section, four N-stage L-Y converter configurations are discussed. To achieve very high output voltage, the multiplier is designed using a diode and capacitor circuitry and attached to two-stage L-Y power converter configurations. These topologies also called L-Y MBC because of the multilevel structure at the output port [222,223].

L-Y MBC converter is designed by combining the N-stage voltage multiplier with the Y converter of L-Y converter, obtain the resultant circuit [222,223]. Thus, L-Y MBC configurations are the hybridization of an L converter and a multistage Y

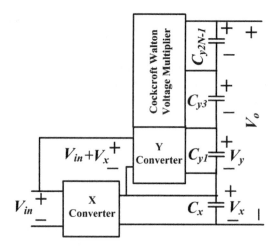

FIGURE 4.15   Universal Structure of N-Stage L-Y Power Converter Configurations.

power converter (Y with N-stage voltage multiplier). The low DC supply is directly connected to the L converter and series connection of the output voltage of the L converter and input voltage fed as input to the multistage Y power converter. The total output voltage generated at the output of the L-Y MBC configuration is equal to the inverting sum of the output voltage of the L converter and the multistage Y power converter as in equation (4.69), where V_L and V_y are the output voltages of the L converter and the multistage Y power converter, respectively. Figure 4.15 depicts the universal structure of N-stage L-Y power converter configurations.

$$\left. \begin{aligned} Vo &= -\left(V_L + V_y\right), \ V_L = VC_x \\ V_y &= VC_{y2N-1} + VC_{y3} + \ ..... \ +VC_{y1} \\ V_o &= -(VC_x + VC_{y2N-1} + VC_{y3} + \ ..... \ +VC_{y1}) \end{aligned} \right\} \qquad (4.69)$$

Based on the Y converter, N-stage L-Y power converter configurations are divided into four categories:

- N-stage L-L power converter configurations
- N-stage L-2L power converter configurations
- N-stage L-2LC power converter configurations
- N-stage L-2LC$_m$ power converter configurations

The power circuit of an N-stage L-L converter configuration is depicted in Figure 4.16(a). An N-stage L-L power converter configuration is also called a L-L multilevel boost converter, which is designed by using an L converter and L-MBC (N-stage L converter or voltage multiplier-based L converter).

A total of 2 N capacitors, 2(0.5 + N) diodes, and 2 inductors, along with a single controlled switch, are required to design an N-stage L-L converter configuration.

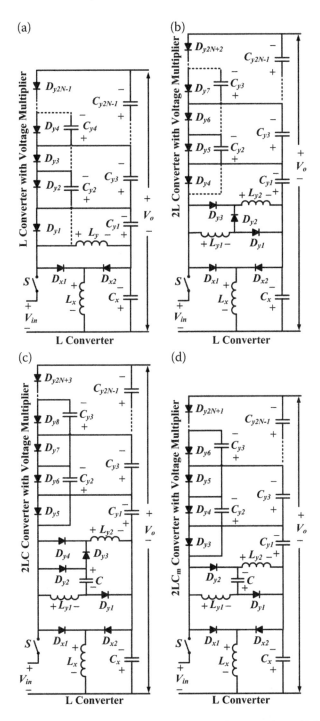

FIGURE 4.16 Power Circuit of N-Stage X-Y Converter Configurations: (a) L-L MBC, (b) L-2L MBC, (c) L-2LC MBC, (d) L-2LC$_m$ MBC.

The power circuit of the N-stage L-2L converter configuration is depicted in Figure 4.16(b). The N-stage L-2L power converter configuration is also called an L-2L multilevel boost converter, which is designed by using an L converter and 2L-MBC (N-stage 2L converter or voltage multiplier-based 2L converter). A total of 2 N capacitors, 2(N + 2) diodes, and 3 inductors, along with a single controlled switch, are required to design an N-stage L-2L converter configuration. The power circuit of the N-stage L-2LC converter configuration is depicted in Figure 4.16(c). An N-stage L-2LC power converter configuration is also called an L-2LC multi-level boost converter, which is designed by using an L-converter and a 2LC-MBC (N-stage 2LC converter or voltage multiplier-based 2LC converter). A total of 2N + 1 capacitors, 2(N + 2.5) diodes, and 3 inductors, along with a single controlled switch, are required to design an N-stage L-2LC converter configuration. The power circuit of the N-stage L-2LC$_m$ converter configuration is depicted in Figure 4.16(d). An N-stage L-2LC$_m$ power converter configuration is also called a L-2LC$_m$ mul-tilevel boost converter, which is designed by using a L converter and a 2LC$_m$-MBC (N-stage 2LC$_m$ converter or voltage multiplier-based 2LC$_m$ converter). A total of a 2N + 1 capacitor, 2(N + 2.5) diodes, and 3 inductors, along with a single controlled switch are required to design an N-stage L-2LC$_m$ converter configuration.

4.5.1 WORKING MODES AND VOLTAGE CONVERSION RATIO OF AN N-STAGE L-Y POWER CONVERTER FAMILY

The working of the N-stage L-Y converter (L-Y MBC converter) configuration is divided into two modes: first, when the switch turned ON (conducting) and another when the switch is turned OFF (not conducting). To explain the working, Figure 4.16(d) depicts the power circuit of the N-stage L-2LC$_m$ converter, and a steady-state operation is assumed as:

• Input supply is pure DC
• All the capacitors and diodes are ideal.
• Component efficiency is 100%; thus, all the components are ideal.
• All the inductors of the X converter are the same and identical in rating.
• All the inductors of the Y converter are the same and identical in rating.
• Operating switching frequency is high to reduce voltage ripple of the capacitor.

The equivalent circuit of the N-stage L-2LC$_m$ converter configuration when switch S is conducting is shown in Figure 4.17(a). Via diode D_{x1} and switch S, the inductor L_x is magnetized by input supply V_{in}. Via switch S and diode D_{y1}, inductor L_{y1} is magnetized by the addition of input supply voltage and the voltage across C_x. Via switch S and diode D_{y2}, inductor L_{y2} is magnetized by the addition of input voltage V_{in} and voltage across C_x. Both inductor L_{y1} and L_{y2} are charged in parallel with capacitor C. At the same time, capacitors C_{y2N-2}, C_{y4}...., and C_{y2} are charged.

The equivalent circuit of the N-stage L-2LC$_m$ converter configuration when switch S is not conducting is shown in Figure 4.17(b). Via diode D_{x2}, inductor L_x is demagnetized and transfers its energy to capacitor C_x. Via diode D_{y3}, capacitor C is discharged in series with inductors L_{y2} and L_{y1} to charge capacitor C_{y1}. Via diode $Dy5$, capacitor $Cy2$

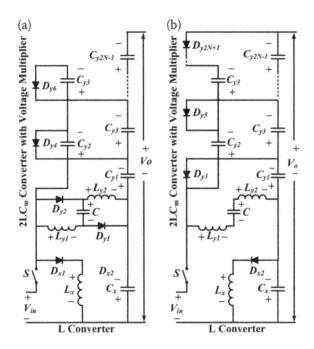

FIGURE 4.17 Power Circuit of L-2LC$_m$ MBC: (a) When Switch S Is Conducting, (b) When Switch S Is Not Conducting.

transfers its energy to charge capacitor C_{y3}. Similarly, capacitors C_{y2N-2}, C_{y6}......, and C_{y4} transferred their energy to charge capacitors C_{y2N-1}, C_{y7}...., and C_{y5}, respectively.

$$G_X = G_{XL} = \frac{V_{Cx}}{V_{in}} = \frac{\Delta}{1 - \Delta} \ or \ V_{Cx} = \frac{\Delta}{1 - \Delta}V_{in} \tag{4.70}$$

$$G_{Y1} = \frac{V_{Cy_1}}{V_{in}} = \frac{(1 + \Delta)(1 + G_{XL})}{1 - \Delta} \ or \ V_{Cy_1} = \frac{(1 + G_{XL})(1 + \Delta)V_{in}}{1 - \Delta} \tag{4.71}$$

$$\left.\begin{aligned} G_{Y2} &= \frac{VC_{y2}}{V_{in}} = (1 + G_{Y1} + G_{XL}) \\ G_{Y3} &= \frac{VC_{y3}}{V_{in}} = (1 + G_{Y1} + G_{XL}) \\ G_{Y5} &= \frac{VC_{y5}}{V_{in}} = (1 + G_{Y1} + G_{XL}) \\ G_{Y2N-1} &= \frac{VC_{y2N-1}}{V_{in}} = (1 + G_{Y1} + G_{XL}) \end{aligned}\right\} \tag{4.72}$$

$$G_Y = G_{2LC_mMBC} = G_{Y1} + G_{Y3} + G_{Y5}..... + G_{Y2N-1} \tag{4.73}$$

$$G_{LY} = G_{L-2LC_mMBC} = \frac{V_o}{V_{in}} = -(G_X + G_Y) \tag{4.74}$$

G_{LY} or G_{L-2LC_mMBC} are the voltage conversion ratios of the N-stage L-2LC$_m$ converter configuration. In the same way, the voltage conversion ratio of the N-stage L-L converter configuration, N-stage L-2L converter configuration, and N-stage L-2LC converter configuration are derived and given in equations (4.75), (4.76), and (4.77), respectively.

$$\left.\begin{array}{l}
G_X = G_{XL} = \Delta/(1 - \Delta) \\
G_{Y1} = \Delta(G_{XL} + 1)/(1 - \Delta) \\
G_{Y2N-1} = ... = G_{Y3} = G_{Y2} = (1 + G_{Y1} + G_{XL}) \\
G_Y = G_{Y2N-1} + G_{Y2N-2} + ... + G_{Y1} \\
V_o/V_{in} = G_{L-LMBC} = G_{LY} = -(G_X + G_Y)
\end{array}\right\} \tag{4.75}$$

$$\left.\begin{array}{l}
G_X = G_{XL} = \Delta/(1 - \Delta) \\
G_{Y1} = 2\Delta(G_{XL} + 1)/(1 - \Delta) \\
G_{Y2N-1} = ... = G_{Y3} = G_{Y2} = (1 + G_{Y1} + G_{XL}) \\
G_Y = G_{Y2N-1} + G_{Y2N-2} + ... + G_{Y3} + G_{Y2} + G_{Y1} \\
V_o/V_{in} = G_{L-2LMBC} = G_{LY} = -(G_X + G_Y)
\end{array}\right\} \tag{4.76}$$

$$\left.\begin{array}{l}
G_X = G_{XL} = \Delta/(1 - \Delta) \\
G_{Y1} = (G_{XL} + 1)(1 + \Delta)/(1 - \Delta) \\
G_{Y2N-1} = ... = G_{Y3} = G_{Y2} = (1 + G_{XL} + G_{Y1}) \\
G_Y = G_{Y2N-1} + G_{Y2N-2} + ... + G_{Y3} + G_{Y2} + G_{Y1} \\
V_o/V_{in} = = G_{L-2LCMBC} = G_{LY} = -(G_X + G_Y)
\end{array}\right\} \tag{4.77}$$

In Figure 4.18, configurations of the N-stage L-Y power converter are compared with each other in terms of voltage conversion ratio.

4.5.2 VALIDATION OF N-STAGE L-Y POWER CONVERTER CONFIGURATIONS

The functionality and concept of all the N-stage L-Y power converter configurations are verified by simulation software Matrix Laboratory 9.0 (R2016a), considering 4 numbers of stages, input supply of 10 V, power of 100 W, a duty cycle of 60%, and a switching frequency of 50 kHz.

Output voltage and input voltage of a four-stage L-L power converter, four-stage L-2L power converter, four-stage L-2LC power converter, and four-stage L-2LC$_m$ power converter are shown in Figure 4.19(a)–(d).

It is investigated that −240 V is achieved from the four-stage L-L power converter. It is investigated that −390 V is achieved from the four-stage L-2L power converter. It is investigated that −490 V is achieved from four-stage L-2LC and L-2LC$_m$ power converters. At a duty cycle of 60%, the voltage conversion ratio of

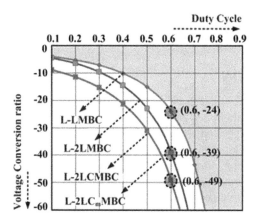

FIGURE 4.18 Comparison of N-Stage L-Y Configuration in Terms of Voltage Conversion Ratio.

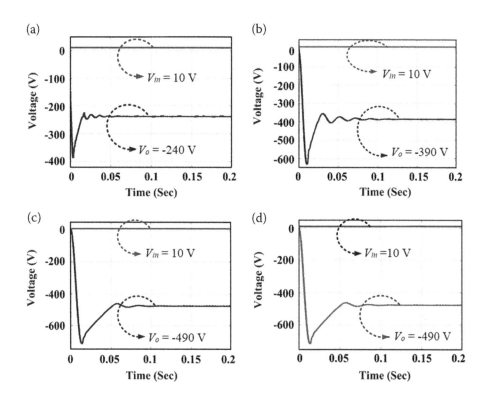

FIGURE 4.19 Simulation Result of N-Stage X-Y Converter Configuration, Output Voltage: (a) L-L MBC, (b) L-2L MBC, (c) L-2LC MBC, (d) L-2LC$_m$ MBC.

four-stage L-L, four-stage L-2L, four-stage L-2LC, and four-stage L-2LC$_m$ converters are -24, -39, -49, and -49, respectively.

4.6 CONCLUSION

A new breed of BBC called the X-Y power converter family is discussed. SI and VLSI play a major role in designing the X-Y converter family. Initially, four BBC configurations are suggested for step-up application by using a SI and VLSI. The XY power converter family is categorized into three groups: two-stage X-Y power converter configuration, three-stage X-Y power converter configuration, and N-stage L-Y power converter configuration. In two-stage X-Y power converters, 16 configurations are explained in detail. In three-stage X-Y power converters, 16 configurations are derived and explained in detail. The three stages are nothing but the addition of a voltage doubler in two-stage X-Y configurations. For an N-stage L-Y power converter, four power converter stages are explained. Further, the N-stage L-Y structures are a combination of a voltage multiplier and a two-stage X-Y converter. Moreover, all the configurations are compared with each other. All of these configurations are a single switch configuration and have a high-voltage conversion ratio. The complete theoretical analysis is provided and the concept is verified by a simulation result. Simulation results always shown a good agreement with the original theoretical hypotheses developed.

5 Self-Balanced DC-DC Multistage Power Converter Configuration without Magnetic Components for Photovoltaic Applications

5.1 INTRODUCTION

Advantages of DC-DC multistage converter configurations are explained in Chapter 3. A self-balanced DC-DC multistage power converter configuration without magnetic components for photovoltaic applications is discussed in this chapter. The multistage converter configuration is derived for photovoltaic applications where high voltage is required, without using magnetic components (without inductor and transformer) [49]. The suggested converter is designed to transfer the power of a photovoltaic source in unidirection with step-up voltage. To get output voltage $N_S + 1$, time is compared to the input voltage; we design the converter configuration with the N_S number of the stage by utilizing a $2N_S$ number of diodes and $2N_S$ number of capacitors. The striking feature of the suggested multistage converter configuration is explained. Detailed theoretical design and explanation of the suggested converter configuration is provided. The effect of semiconductor losses on a voltage conversion ratio of the suggested converter configuration is provided. We discuss the complete theoretical background and design equation for the rating of components of the suggested converter. Recent multistage converter configuration without inductor and transformer is also discussed with advantages and disadvantages, compared with the suggested configuration in terms of cost, voltage conversion ratio, and the number of semiconductor devices and capacitors. The functionality and concept of suggested converter configurations verified by using by the simulation software Matrix Laboratory 9.0 (R2016a) environment. We discuss the hardware implementation by using the PIC controller and hardware results show a good agreement with theoretical and simulation result.

---

*$V_{out} = V_o$

5.2 RECENT DC-DC MULTISTAGE CONVERTER WITHOUT INDUCTOR AND TRANSFORMER

In this section, a recently addressed DC-DC multistage converter configuration without an inductor and transformer is reviewed and their problem explained [26–33,49]. In these configurations, capacitors are highly used as boosting elements, and charging and discharging of capacitors are controlled through semiconductor switches.

5.2.1 DC-DC MULTISTAGE FLYING CAPACITOR CONVERTER (M-FCC) CONFIGURATION

The power circuit of DC-DC three-stage flying capacitor converter circuit (M-FCC) depicted in Figure 5.1. This converter configuration is bidirectional and operates in a buck and boost mode [49,224–226]. In buck mode operation, this configuration provides output voltage $1/N_S$ times compared to the input voltage. In boost mode operation, this configuration provides output voltage N_S times compared to the input voltage. It reports that the efficiency of the converter in the literature is greater than 95%, and the voltage across all the switches is equal [224–226].

On the other hand, M-FCC has the following drawbacks:

- The converter does not have a modular structure. Thus, the configuration is not readily extendable to a more significant number of stages.
- Converter configuration size is large because of many capacitors and transistor required for the design. To design the N_S stage FCC, it requires N_S number of capacitors and $2N_S$ number of switches to design converter configuration.
- It requires a complicated switching scheme to operate the converter.
- It is not possible to operate at a high switching frequency when the turn ON time of the switch is comparable to the fall and rise time. At high frequency, the energy of the input source is transferred to the capacitors in a short time; therefore, the efficiency of the converter is low.

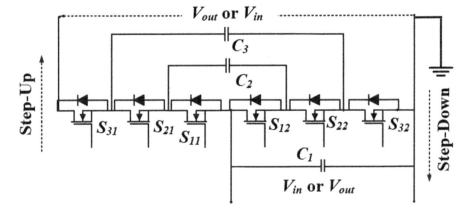

FIGURE 5.1   Power Circuit of Three-Stage Flying Capacitor Converter Configuration.

- If any control switch does not fail, then it is not possible to control the power circuit.
- Utilization of components is less.

5.2.2 DC-DC Multistage Switched Series-Parallel Capacitor Converter (SSPCC) Configuration

The power circuit of three-stage switched series-parallel capacitor converter (SSPCC) configuration is shown in Figure 5.2. This configuration is designed by using only switches and capacitors [49,226–228].

The operation of this converter is divided into two modes. All the control switches are operated in such a way that all the capacitors are discharged in parallel and charge in series. This converter configuration is bidirectional and operates in a buck and boost mode. In buck mode operation, this configuration provides output voltage $1/N_S$ times compared to the input voltage. In boost mode operation, this configuration provides output voltage N_S times compared to the input voltage. It reports that the efficiency of the converter in the literature is over 90%. Nevertheless, SSPCC has the following drawbacks:

- Because of many switching elements, it is complicated to change switching states.
- It requires a various rating of power switches because of the unequal voltage across each switch.
- Though converter configuration is bidirectional but difficult to control the direction of power flow, the direction depends upon the input and output voltages of the DC bus.
- An unbalanced voltage situation may occur in the capacitors if the proper control scheme is not applied.

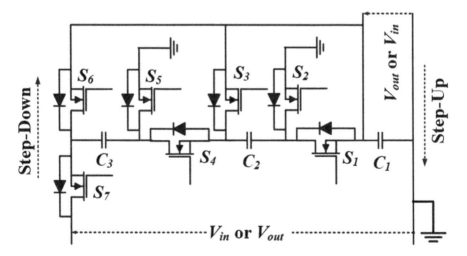

FIGURE 5.2 Power Circuit of Three-Stage Switched Series-Parallel Capacitor Converter (SSPCC) Configuration.

- Converter configuration is not modular because of the requirement of a large number of diodes, switches, and gate driver increases with several stages. Thus, the converter is a larger size and has a higher weight.

5.2.3 DC-DC Multistage Fibonacci Converter (MFC) Configuration

The power circuit of the DC-DC three-stage Fibonacci converter configuration is shown in Figure 5.3. The configuration derived by using switches and capacitors circuitry and the Fibonacci series is shown by the voltage conversion ratio [49,229–231].

At the output side we achieve high voltage; this configuration required a large number of capacitors and switches. However, a DC-DC MFC configuration has the following drawbacks:

- It requires $3N_s + 1$ number of diodes and switches to design an N_s stage Fibonacci converter configuration. Thus, it required many diodes and switches.
- The converter configuration is unidirectional.
- The voltage conversion ratio of converters follows Fibonacci series and hence it is impossible to attain a conversion ratio, which is not in a Fibonacci series like 2, 4....................

5.2.4 DC-DC Multistage Magnetic-Free Converter (MMC) Configuration

The power circuit of the DC-DC three-stage magnetic-free converter configuration is shown in Figure 5.4. The converter configuration is designed by connecting a modular block in a mesh pattern [49,232]. The modular block of the converter is formed by using one capacitor and two transistors. This converter configuration is bidirectional and operates in a buck and boost mode. In buck mode operation, N_S-stage magnetic-free converter configuration provides output voltage $1/N_S$ times

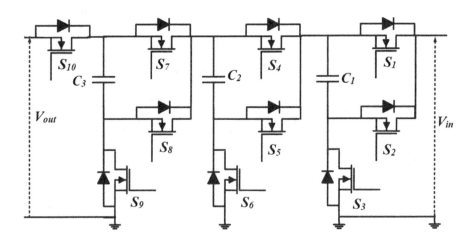

FIGURE 5.3 Power Circuit of DC-DC Three-Stage Fibonacci Converter Configuration.

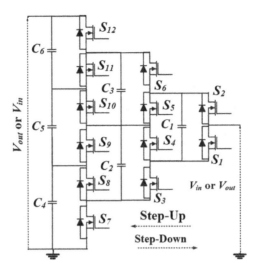

FIGURE 5.4   Power Circuit of the DC-DC Three-Stage Magnetic-Free Converter Configuration.

compared to the input voltage. In boost mode operation, N_S-stage magnetic-free converter configuration provides output voltage N_S times compared to the input voltage.

The voltage across each transistor is identical and equal to input supply; thus, it is independent of the voltage conversion ratio and duty cycle. The DC-DC MMC configuration has the following drawbacks:

- To design the DC-DC MMC configuration for the N_S number of stages, $N_S \times (N_S + 1)$ number of control switches, $0.5 \times (N_S \times 2 + N_s)$ number of capacitors, and $N_S (N_S + 1)$ number of diodes are required. Hence, many switching devices and capacitor are required, which increases the size of the converter.
- It is challenging to manage the power flow direction of the converter because many switches are available for the conducting path. Hence, the configuration also required complex circuitry.
- The direction of the power flow of the converter depends on the terminal voltage at DC-Bus. Thus, this converter is not suitable for the application where input source voltage is varying in characteristics.

5.2.5   DC-DC Step-Up Modified Switched-Mode Converter Configuration or DC-DC Switched-Mode Converter Configuration

The power circuit of the DC-DC step-up modified switched-mode converter configuration or DC-DC switched-mode converter configuration is shown in Figure 5.5. The converter configuration is simple in control, capable of offering continuous input current, and the voltage conversion ratio is adjusted by changing the ON time switches [49,233].

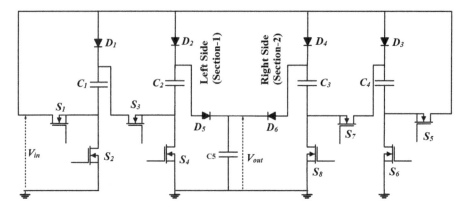

FIGURE 5.5 Power Circuit of DC-DC Step-Up Modified Switched-Mode Converter Configuration or DC-DC Switched-Mode Converter Configuration.

In Table 5.1, the detailed study about the working modes of converter and capacitor states are provided. The Electro Magnetic Interference (EMI) problem is reduced because of continuous current from the low-voltage source at the input side. The DC-DC switched-mode converter configuration has the following drawback:

- It requires more switches.
- Efficiency is low because switching losses are high.
- They do not increase the voltage conversion ratio further because a similar extension is not possible for this configuration.
- The configuration does not provide a good agreement for a high-voltage high-power application due to a low-voltage conversion ratio.

TABLE 5.1

State of Capacitor and Working Mode of Operation of DC-DC Switched-Mode Converter Configuration [49,233]

Working Modes	Switches State								Capacitors of Left Part (Section-1)	Capacitors of Right Part (Section-2)
	S1	S2	S3	S4	S5	S6	S7	S8		
I	#	$	#	$	$	#	$	#	C	D
II	#	#	#	#	$	#	$	#	NA	D
III	$	#	$	#	#	$	#	$	D	C
IV	$	#	$	#	#	#	#	#	D	NA

C: Charging, NA: No Action, D: Discharging, #: OFF state, $: ON state.

5.2.6   DC-DC Switched-Capacitor Converter Configuration

The power circuit of the DC-DC switched-capacitor converter configuration is shown in Figure 5.6. The input current ripple of this configuration is low. The EMI effect is reduced because of continuous input current with low ripples [49,234]. However, there is no provision to increase the voltage conversion ratio.

5.2.7   DC-DC Capacitor Clamped Modular Multilevel Converter (CCMMC) Configuration

The power circuit of the DC-DC capacitor clamped modular five-level converter configuration is shown in Figure 5.7. This capacitor is designed only by using

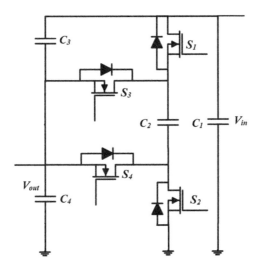

FIGURE 5.6   Power Circuit of the DC-DC Switched-Capacitor Converter Configuration.

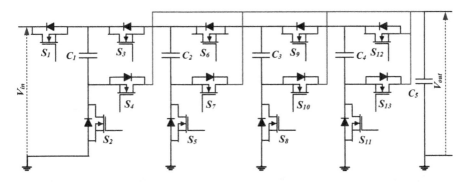

FIGURE 5.7   Power Circuit of the DC-DC Capacitor Clamped Modular Five-Level Converter Configuration.

switches and capacitors [49,235]. The following are the major drawbacks of a CCMMC configuration:

- High voltage is achieved at the output. However, many capacitors and switches are used to design CCMCC configurations.
- It requires high-voltage rating switches due to high voltage stress across the switches. For N_s-level configuration, the voltage stress across N_s–2 switching devices is double the input voltage, and the remaining switches have voltage stress equal to the input voltage.

5.3 SELF-BALANCED DC-DC MULTISTAGE POWER CONVERTER CONFIGURATION WITHOUT MAGNETIC COMPONENTS

In this section, a self-balanced DC-DC multistage power converter configuration without magnetic components is suggested (also called a suggested converter) to overcome the drawbacks of a central inverter photovoltaic configuration (CI-PVC), string inverter photovoltaic configuration (SI-PVC), AC module photovoltaic configuration (ACM-PVC), multistring inverter photovoltaic configuration (MSI-PVC), and the previously discussed converter configuration. For the existing photovoltaic system, the suggested converter configuration provides a practical solution to boost the voltage without an inductor and transformer before feeding energy to a DC-AC converter for a grid or to drive the motor. Figure 5.8 shows the photovoltaic system, where the suggested converter suffices to deliver energy to the DC-AC multilevel converter [49].

To transfer maximum power to the load from the photovoltaic module, it is essential to match the load resistance, R_L, to the maximum output resistance of the photovoltaic module, R_{PV} ($R_{mpp} = I_{mpp}^{-1}V_{mpp}$). Figure 5.9(a) depicts the general

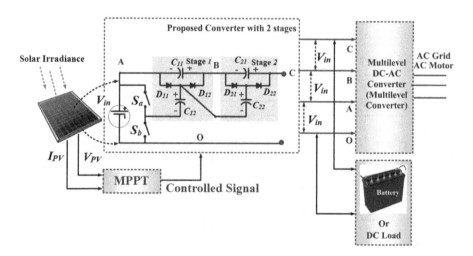

FIGURE 5.8 Self-Balanced DC-DC Multistage Power Converter Configuration without Magnetic Components in a Photovoltaic System to Deliver the Photovoltaic Energy to a Grid, DC Load, or Motor.

characteristic plot of power voltage and current voltage of the photovoltaic module. Zero power is extracted from the photovoltaic module when the photovoltaic current (I_{PV}) and voltage (V_{PV}) is equal to the short-circuit current (I_{SC}) and open-circuit voltage (V_{OC}), respectively. By regulating the working voltage of V_{PV}, it is feasible to track the photovoltaic module maximum power point (MPP). In literature, various algorithms are suggested for tracking the MPP to transfer maximum power. Because of simple logic and faster response, perturbation and observation (P&O) is a popular algorithm to track the MPP. A reconfigurable switched-capacitor power converter configuration is suggested, along with tracking the MPP, and a P&O algorithm is discussed for DC-DC converters connected to photovoltaic generators [49]. The idea to manage the maximum power of a self-balanced DC-DC multistage power converter configuration is explained in Figure 5.9(b)–(f)[49]. To control the working voltage, V_{PV}, in the photovoltaic system, the number of stages/levels (if the configuration is reconfigurable) and the capacitor charging or discharging time are two controlled parameters. By controlling these parameters

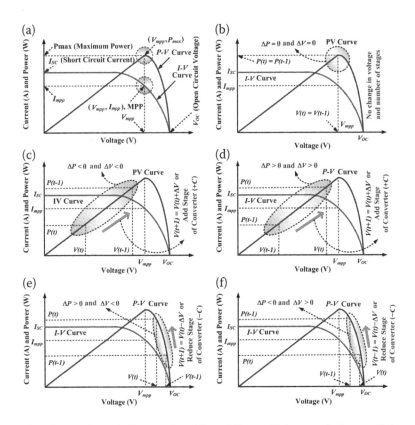

FIGURE 5.9 (a) General Characteristic Plot of Power Voltage and Current Voltage of Photovoltaic Module, (b) Tracking MPP When $\Delta V = \Delta P = 0$, (c) Tracking MPP When $\Delta V < 0$ and $\Delta P < 0$, (d) Tracking MPP When ΔV and $\Delta P > 0$, (e) Tracking MPP When $0 < \Delta P$ and $0 > \Delta V$, (f) Tracking MPP When $0 > \Delta P$ and $0 < \Delta V$.

through MPPT algorithms, the suggested converter system is operating at the MPP to extract maximum power from the photovoltaic source.

The power circuit of the self-balanced DC-DC multistage power converter configuration without magnetic components for K stages is shown in Figure 5.10. The striking features of the presented converter configuration include the following:

- The power circuit is transformer-less and inductor-less.
- The input current is continuous.
- The rating of the capacitors and semiconductor device is low.
- The converter configuration is modular.
- It is trouble-free to add more levels to attain a higher voltage conversion ratio.
- The number of control switches is only two, which operate alternatively.
- The control of the converter is simple.

The multistage converter configuration is applicable where high voltage is required without using magnetic components (without inductor and transformer) [49,224–235]. The self-balanced converter (Figure 5.10) achieves higher output voltage without utilizing a magnetic component and to transfer the power of the photovoltaic source in unidirection with step-up voltage. High frequency can drive the switches, which allow the reduction in the capacitor size, and thus the converter is reduced.

The number of switches and its control is independent of the number of stages. However, the number of capacitors and diodes depends on the number of stages. Two capacitors and two diodes are necessary to raise the number of stages of the suggested configuration by one. For example, 8 capacitors, 8 diodes (uncontrolled switches), and 2 switches are required to design four stages, as suggested in the converter configuration.

To explain the operation modes of the magnetic component-free K stages converter (Figure 5.10) circuit considered. The mode of operation of the magnetic component-free K stages converter is divided into two modes: mode 1 when switches S_b and S_a act as a short circuit (turned ON) and open circuit (turned OFF), respectively, and mode 2 when switches S_a and S_b act as a short circuit (turned ON) and open circuit (turned OFF), respectively. Hence, switches S_a and S_b are

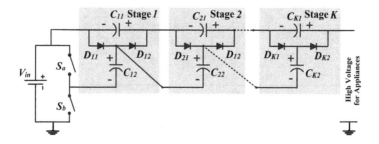

FIGURE 5.10 Self-Balanced DC-DC Multistage Power Converter Configuration without Magnetic Components for K Stages.

complementary in operation. The magnetic component-free K stages converter has a simple control and is operated at a fixed duty cycle of 0.5% to provide voltage to photovoltaic devices. A complicated gate driver is also not required to drive the switch; instead, an oscillator is sufficient to provide a gated signal.

When switch S_a is turned OFF and switch S_b is turned ON, input supply voltage V_{in} charges the C_{12} capacitor via diode D_{11} and switch S_b when the input supply voltage is greater than the voltage across capacitor C_{12}. The capacitor C_{22} is charged by capacitor C_{11} through diode D_{21} and switch S_b when voltage $V_{in} + VC_{11}$ is more significant than voltage $VC_{12} + VC_{22}$. Similarly, the capacitor C_{K2} is charged by capacitor $C_{(k-1)1}$ through diode D_{K1} and switch S_b, when voltage $V_{in} + VC_{11} + VC_{21} + ... + VC_{(k-1)1}$ is more significant than voltage $VC_{12} + VC_{22} + VC_{32} + .. + VC_{k2}$. The output voltage is the addition of input supply voltage, the voltage across capacitor C_{11}, voltage across capacitor C_{21}, ..., and the voltage across capacitor C_{K1}. The equivalent circuit of the converter when switch S_b is turned ON is shown in Figure 5.11.

When switch S_b is turned OFF and switch S_a is turned ON, capacitor C_{12} transfers its energy to charge capacitor C_{11} via diode D_{12} and switch S_a when the voltage across capacitor C_{12} is greater than the voltage across capacitor C_{11}. The capacitor C_{21} is charged by capacitor C_{22} via diode D_{22} and switch S_a when the voltage across capacitor $C_{12} + C_{22}$ is greater than the voltage across capacitor $C_{11} + C_{21}$. Similarly, if the voltage across capacitor $C_{12} + C_{22} + C_{K2}$ is greater than the voltage across capacitor $C_{11} + C_{21} + C_{K1}$, then capacitor C_{K1} is charged by the capacitor C_{K2} via diode D_{K2}. The output voltage is the addition of input supply voltage (V_{in}), voltage across capacitor C_{12}, voltage across capacitor C_{22}...., and the voltage across capacitor C_{K2}. We show the equivalent circuit of the converter when switch S_a is turned ON in Figure 5.12.

5.4 VOLTAGE CONVERSION RATIO ANALYSIS WITHOUT CONSIDERING THE LOSS IN SWITCHES AND DIODES

If the voltage drop across all the switches and diodes is not considered, then the voltage across all the capacitors is equal to the input supply voltage (V_{in}). The voltage conversion ratio of a K-stage self-balanced power converter configuration is $K + 1$. Thus, the voltage conversion ratio depends on the number of stages or capacitors. The

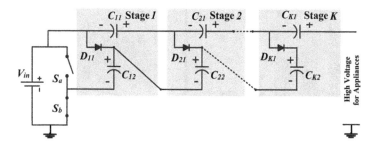

FIGURE 5.11 Equivalent Circuit of Self-Balanced DC-DC Multistage Power Converter Configuration without Magnetic Components for K Stages When Switch S_b Is Turned ON.

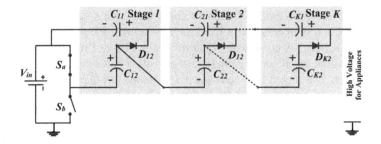

FIGURE 5.12 Equivalent Circuit of Self-Balanced DC-DC Multistage Power Converter Configuration without Magnetic Components for K Stages When Switch S_a Is Turned ON.

graph of the number of components or devices versus stages of the converter is shown two-dimensionally in Figure 5.13(a), and for better understanding, the graph is also shown three-dimensionally in Figure 5.13(b).

We see that the number of components and devices is linearly increasing with the number of stages. Two diodes and capacitors are required to design one stage of the converter. In this configuration, the number of capacitors and the number of diodes is analyzed and the characteristic equation as follows:

$$V_{in} = V_{CK2} = V_{C22} = V_{C21} = V_{C11} = V_{C12} \tag{5.1}$$

$$V_o = V_{in} \times (1 + K) \tag{5.2}$$

$$V_o = V_{in} \times \frac{1}{2} \times (2 + K_C) \tag{5.3}$$

$$V_o = V_{in} \times \frac{1}{2} \times (2 + K_D) \tag{5.4}$$

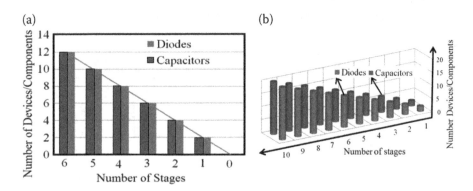

FIGURE 5.13 Graph of the Number of Components or Devices Versus the Number of Stages: (a) Two-Dimensional Graph, (b) Three-Dimensional Graph.

$$K_D = K_C = \frac{K}{2} \tag{5.5}$$

The number of diodes is K_D and the number of the capacitors is K_C when used to design the converter. Figure 5.14(a) depicts the plot of the voltage conversion ratio and the number of stages. We see that the $1 + K$ voltage conversion ratio is achieved by a K-stage converter configuration. Figure 5.14(b) depicts the plot of the number of components/devices and voltage conversion ratio. We observe that the requirement of the number of components or devices linearly increases. Also, it is noted that the number of devices/components linearly increases as the voltage gain requirement increases. Thus, two diodes and two capacitors are needed to be connected to increase the voltage gain by 1. It is also observed that $2K$ diodes and $2K$ capacitors are required to attain a voltage gain of $K+1$. The plot of the number of stages, voltage conversion ratio, and duty cycle (D or Δ) is shown in Figure 5.15(a). Note that two diodes and two capacitors are required to design a single stage of the converter. The plot of the output voltage, assuming a 25-V input for stages 1 to 9, is shown in Figure 5.15(b) and we observe that the output voltage is rising with the number of stages. It is also seen that one stage of the converter has contributed a voltage equal to the input voltage (25 V) in output voltage.

5.5 VOLTAGE CONVERSION RATIO ANALYSIS WHEN CONSIDERING THE LOSS OF SWITCHES AND DIODES

The power diode and switches voltage drop is not considered in high- and medium-voltage applications, but is in low-voltage applications, as discussed in this section. The suggested converter configuration is studied by assuming the power diode and switch drop. To avoid complicated equations, and for simplicity, V_d is considered a power diode and switch voltage drop. The voltage across capacitors is provided in equation (5.6).

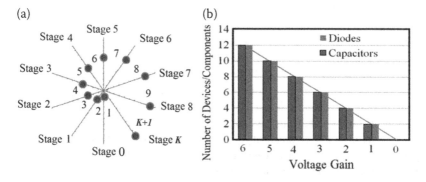

FIGURE 5.14 (a) Two-Dimensional Graph of Voltage Conversion Ratio for Many Stages, (b) Three-Dimensional Graph of Number of Components/Devices for a Voltage Conversion Ratio.

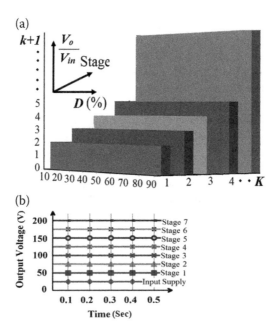

FIGURE 5.15 (a) Graph of Voltage Conversion Ratio, Duty Cycle, and Number of Stages, (b) Graph of Output Voltage for Stages 1 to 7 When Considering Input Voltage (25 V).

$$
\left.
\begin{aligned}
V_{C11} &= -V_{D12} + V_{C12} - V_{Sa} \\
V_{C11} &= V_{in} - 4V_d \\
V_{C12} &= V_{in} - 2V_d \\
V_{C21} &= V_{C12} - V_{C11} - V_{Sa} + V_{C22} - V_{D22} = V_{in} - 4V_d \\
V_{C22} &= V_{in} - V_{C12} + V_{C11} - V_{Sb} - V_{D21} = V_{in} - 4V_d \\
V_{CK1} &= V_{in} - 4V_d \\
V_{CK2} &= V_{in} - 4V_d
\end{aligned}
\right\}
\tag{5.6}
$$

We observed that $V_{in} - 4V_d$ is the potential across each capacitor, except the potential across C_{12} capacitor. $V_{in} - 2V_d$ is the potential across C_{12} capacitor. For this reason, the suggested converter configuration is self-balanced. However, the number of devices and its forward voltage restricts the voltage conversion ratio (duty ratio) of the suggested converter. When considering $V_d = 1$ (practical diode) and $V_d = 0$ (ideal diode), the plot of the output voltage to the number of stages is shown in Figure 5.16(a). Also, when considering $V_d = 1$ (practical diode) and $V_d = 0$ (ideal diode), the plot of the output voltage to the number of capacitors or diodes is shown in Figure 5.16(b).

We can see that the difference between output voltage with $V_d = 1$ (practical diode) and $V_d = 0$ (ideal diode) increases with the number of stages and the number of capacitors and diodes. This differs based on the number of stages and voltage drops across the diode, and it is equal to $4V_d K$, as provided in equation (5.7).

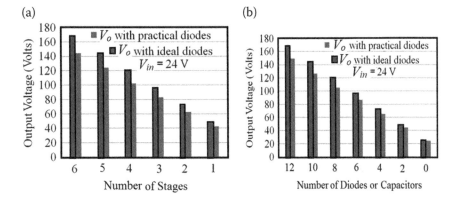

FIGURE 5.16 Plot When Considering 24-V Input Supply, $V_d = 1$ (Practical Diode) and $V_d = 0$ (Ideal Diode): (a) Plot of Output Voltage to Number of Stages, (b) Plot of Output Voltage Concerning Number of Capacitors or Diodes.

$$
\left.
\begin{aligned}
V_{in} - 4V_d &= V_{C11} = V_{C22} = V_{C21} = V_{CK1} \\
V_{in} - 2V_d &= V_{C12} \\
V_O &= -4KV_d + (K + 1)V_{in}
\end{aligned}
\right\}
\qquad (5.7)
$$

5.6 SELECTION OF CAPACITOR

The suggested converter configuration is considered with a single stage to explain the selection of capacitors. Figure 5.17 depicts the power circuit of the suggested converter with a single stage.

The equivalent circuit of the converter when S_b is turned ON, and S_a is turned ON is depicted in Figure 5.18(a) and (b), respectively, where forward diode resistance is R_D, forward switch resistance is R_S, the current flowing through switch S_a is I_{Sa}, and current flowing through switch S_b is I_{Sb}. The capacitor C_{12} voltage is not instantaneously increased to the input voltage V_{in}; the voltage is exponentially built,

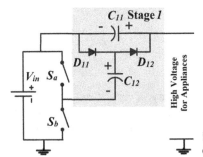

FIGURE 5.17 Power Circuit of Suggested Converter Configuration with a Single Stage.

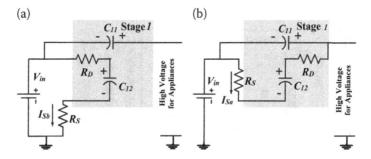

FIGURE 5.18 Equivalent Power Circuit of Suggested Converter: (a) When Switch S_b Is Conducting, (b) When Switch S_a Is Conducting.

not linearly. The numerical equations can be obtained as given in equations (5.8)–(5.12).

$$
\left.
\begin{aligned}
V_{in} &= v_{C_{12}} + (R_D + R_S)\frac{d\,(v_{C12})\,C_{12}}{dt} \\
\frac{d\,(v_{C12})}{dt} &= \frac{V_{in} - v_{C12}}{C_{12}(R_D + R_S)}
\end{aligned}
\right\}
\tag{5.8}
$$

$$
\left.
\begin{aligned}
\int \frac{d\,(V_{C12})}{-V_{C12} + V_{in}} &= \int \frac{dt}{C_{12}(R_S + R_D)} \\
\log(V_{in} - V_{C12}) &= K + \frac{-t}{C_{12}(R_S + R_D)}, \\
K &= \log V_{in} \\
V_{C12} &= \left(-e^{\frac{-t}{T}} + 1\right)V_{in},\ T = C_{12}(R_S + R_D)
\end{aligned}
\right\}
\tag{5.9}
$$

$$
i_{C_{12}} = \frac{d\,(v_{C12})}{dt}C_{12} = \frac{d\,(v_{C12}\,C_{12})}{dt}
\tag{5.10}
$$

$$
V_{in} = V_{C_{12}} + (R_S + R_D)i_{C_{12}}
\tag{5.11}
$$

Similarly,

$$
i_{Sb} = e^{\frac{-t}{T}}\frac{V_{in}}{(R_S + R_D)}
\tag{5.12}
$$

When switch S_a is turned ON, capacitor C_{12} charges capacitor C_{11} through resistance R_S and R_D. Therefore, in this state, capacitor C_{12} is discharging and capacitor C_{11} is charging. The numerical equations are given in equations (5.13)–(5.15).

$$V_{C_{l2}} = V_{C_{l1}} + (R_S + R_D)i_{C_{l2}} \qquad (5.13)$$

$$\left. \begin{aligned} V_{C_{l2}} &= V_{C_{l1}} + (R_S + R_D)\frac{d(v_{C_{l1}})}{dt}C_{l1} \\ \frac{d(v_{C_{l1}})}{V_{C_{l2}} - V_{C_{l1}}} &= \frac{dt}{C_{l1}(R_S + R_D)} \end{aligned} \right\} \qquad (5.14)$$

$$\left. \begin{aligned} \int \frac{d(v_{C_{l1}})}{V_{C_{l2}} - V_{C_{l1}}} &= \int \frac{dt}{C_{l1}(R_S + R_D)} \\ \log(V_{C_{l2}} - V_{C_{l1}}) &= K + \frac{-t}{C_{l1}(R_S + R_D)} \\ K &= \log V_{C_{l2}} \\ V_{C_{l1}} &= \left(1 - e^{\frac{-t}{T}}\right)V_{C_{l2}}, \; T = C_{l1}(R_S + R_D) \end{aligned} \right\} \qquad (5.15)$$

Similarly, with a high switching frequency and in steady state, at any instant the voltage across capacitors C_{12} and C_{11} is calculated by equations (5.16) and (5.17), where $V_{C'12}$ and $V_{C'11}$ are the initial voltages across capacitors C_{12} and C_{11}.

If the initial capacitor voltage is positive, then

$$\left. \begin{aligned} V_{C_{l2}} &= V_{C'_{12}} + \left(-e^{\frac{-t}{T}} + 1\right)\left(-V_{C'_{12}} + V_{in}\right) \\ V_{C_{l1}} &= V_{C'_{11}} + \left(-V_{C'_{11}} + V_{C_{l2}}\right)\left(-e^{\frac{-t}{T}} + 1\right) \end{aligned} \right\} \qquad (5.16)$$

If the initial capacitor voltage is negative, then

$$\left. \begin{aligned} V_{C_{l2}} &= \left(-e^{\frac{-t}{T}} + 1\right)\left(V_{C'_{12}} + V_{in}\right) - V_{C'_{12}} \\ V_{C_{l1}} &= \left(-e^{\frac{-t}{T}} + 1\right)\left(V_{C'_{11}} + V_{C_{l2}}\right) - V_{C'_{12}} \end{aligned} \right\} \qquad (5.17)$$

The required time for the C_{12} capacitor to reach any voltage value throughout the charging cycle is provided in equations (5.18) and (5.19).

If the initial capacitor C_{12} voltage is positive, then

$$t = T\log\left(\frac{-V_{C'_{12}} + V_{in}}{-V_{C_{l2}} + V_{in}}\right) = C_{12}(R_S + R_D)\log\left(\frac{-V_{C'_{12}} + V_{in}}{-V_{C_{l2}} + V_{in}}\right) \qquad (5.18)$$

If the initial capacitor C_{12} voltage is negative, then

$$t = T\log\left(\frac{V_{C'_{12}} + V_{in}}{-V_{C_{l2}} + V_{in}}\right) = C_{12}(R_S + R_D)\log\left(\frac{V_{C'_{12}} + V_{in}}{-V_{C_{l2}} + V_{in}}\right) \qquad (5.19)$$

The required time for the C_{11} capacitor to reach any voltage value throughout the charging cycle is provided in equations (5.20) and (5.21).

If the initial capacitor C_{11} voltage is positive, then

$$t = Tlog\left(\frac{-V_{C'11} + V_{C_{12}}}{-V_{C_{11}} + V_{C_{12}}}\right) = C_{11}(R_S + R_D)\log\left(\frac{-V_{C'11} + V_{C_{12}}}{-V_{C_{11}} + V_{C_{12}}}\right) \qquad (5.20)$$

If the initial capacitor C_{11} voltage is negative, then

$$t = Tlog\left(\frac{V_{C'11} + V_{C_{12}}}{-V_{C_{11}} + V_{C_{12}}}\right) = C_{11}(R_S + R_D)\log\left(\frac{V_{C'11} + V_{C_{12}}}{-V_{C_{11}} + V_{C_{12}}}\right) \qquad (5.21)$$

$$\left.\begin{aligned} C_{12} &= \frac{1}{X_{C12}2\pi f_s} = \frac{1}{\frac{V_{C12}}{I_{C12}}2\pi f_s} = \frac{I_{C12}}{V_{C12}2\pi f_s} \\ C_{11} &= \frac{1}{X_{C11}2\pi f_s} = \frac{1}{\frac{V_{C11}}{I_{C11}}2\pi f_s} = \frac{I_{C11}}{V_{C11}2\pi f_s} \end{aligned}\right\} \qquad (5.22)$$

During the entire switching cycle, the voltage across all the capacitors is the same and equal to the input supply. Thus, the same rating capacitors are required to design the suggested converter.

5.7 COMPARISON OF CONVERTER CONFIGURATIONS

In Table 5.2, the suggested converter configurations are compared with previously discussed transformer-less and inductor-less converter configurations in terms of the number of controlled switches.

TABLE 5.2

Comparison of Converter Configurations in Terms of the Number of Controlled Switches [49]

Converter Type	Number of Levels/Stages					
	1	2	3	4	5	N
SSPCC	1	4	7	10	13	$3N - 2$
M-FCC	2	4	6	8	10	$2N$
MMC	2	6	12	20	30	$N(N + 1)$
MFC	4	7	10	13	16	$3N + 1$
Switch Mode	4	8	12	16	20	$4N$
CCMMC	1	4	7	10	13	$3N - 2$
Figure 5.10	2	2	2	2	2	2

We have suggested that a converter configuration needs a smaller number of switches when compared to the recent magnetic component-free converter. The main advantage of the suggested converter is that the number of switches does not depend on the number of stages. Only two control switches are required to design the suggested converter. The suggested converter configuration is compared with previously discussed transformer-less and inductor-less converter configurations in terms of several diodes and capacitors in Tables 5.3 and 5.4, respectively. The suggested converter configuration is compared with previously discussed transformer-less and inductor-less converter configurations in terms of the voltage conversion ratio in Table 5.5. Figure 5.19(a)–(d) depicts the comparison plot of the suggested converter with the recent converter in terms of number of capacitors,

TABLE 5.3

Comparison of Converter Configurations in Terms of the Number of Diodes [49]

Converter Type	Number of Levels/Stages					
	1	2	3	4	5	N
SSPCC	1	4	7	10	13	$3N - 2$
M-FCC	2	4	6	8	10	$2N$
MMC	2	6	12	20	30	$N(N + 1)$
MFC	4	7	10	13	16	$3N + 1$
Switch Mode	4	6	8	10	12	$2N + 2$
CCMMC	1	4	7	10	13	$3N - 2$
Figure 5.10	2	4	6	8	10	$2N$

TABLE 5.4

Comparison of Converter Configurations in Terms of the Number of Capacitors [49]

Converter Type	Number of Levels/Stages					
	1	2	3	4	5	N
SSPCC	1	2	3	4	5	N
M-FCC	1	2	3	4	5	N
MMC	1	3	6	10	15	$N(N + 1)/2$
MFC	1	2	3	4	5	N
Switch Mode	3	5	7	9	11	$2N + 1$
CCMMC	1	2	3	4	5	N
Figure 5.10	2	4	6	8	10	$2N$

TABLE 5.5

Comparison of Converter Configurations in Terms of theVoltage Conversion Ratio [49]

Converter Type	Number of Levels/Stages					
	1	2	3	4	5	N
SSPCC	1	2	3	4	5	N
M-FCC	1	2	3	4	5	N
MMC	1	2	3	4	5	N
MFC	1	3	5	8	Not feasible to design for higher levels	
Switch Mode	2	3	4	5	6	N + 1
CCMMC	1	2	3	4	5	N
Figure 5.10	2	3	4	5	6	N + 1

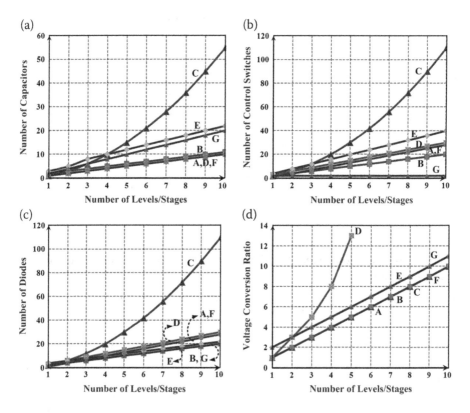

FIGURE 5.19 Comparison Plot of Suggested Converter with Previously Discussed Converter: (a) In Terms of Several Capacitors, (b) In Terms of Number of Switches, (c) In Terms of Several Diodes, (d) In Terms of Voltage Conversion Ratio (Suggested Converter= G, CCMMC=F, Switch Mode=E, MFC=D, MMC=C, M-FCC=B, SSPCC=A).

number of switches, number of diodes, and voltage conversion ratio. Using a graph we find that the suggested converter configuration offers the best solution in terms of several components. In terms of cost, the suggested converter is compared with recent converters and the comparison is provided in Table 5.6. We find that only two switches are required to design the suggested converter configuration. Therefore, the cost of the suggested converter configuration is less when compared to other recent converters.

5.8 VALIDATION OF SELF-BALANCED DC-DC MULTISTAGE POWER CONVERTER

In this section, the experimental and simulation result of a self-balanced DC-DC multistage power converter configuration without magnetic components for four stages is presented and discussed in detail. The suggested converter configuration is designed with 60-W-rated power, 24-V input supply, 100-kHz switching frequency, and four stages. To reduce capacitor rating and size, a high switching frequency is used. Both switches S_a and S_b are controlled complementarily with a duty cycle of 50%.

Figure 5.20(a) depicts the current and voltage waveform without considering diode and switch voltage drop. We find that without considering the diode and switch voltage drop, the output voltage settles at 2 ms. The effect of a diode and switch drop is notable, and Figure 5.20(b) depicts the current and voltage waveform when considering a diode and switch voltage drop equal to 1 V. The output voltage settles at 4 ms when considering the 1-V diode and switch voltage drop. Due to the time constant $(R_S+R_D)C$, the output waveform in both cases differs. The power of the converter and voltage across the switch is shown in Figure 5.20(c)–(d), respectively.

The input and output voltage waveform with or without considering the voltage drop across devices is shown in Figure 5.20(e)–(f), respectively. Without

---

TABLE 5.6

Comparison of Converter Configurations in Terms of the Cost of the Converter

Converter Type	Cost of the Converter (N Is Number of Stages)
SSPCC	$(3N - 2)$ (cost of each switch + cost of each diode) + N (cost of each capacitor)
M-FCC	$2N$ (cost of each switch + cost of each diode) + N (cost of each capacitor)
MMC	$N(N + 1)\{$(cost of each switch + cost of each diode) + 0.5 (cost of each capacitor)$\}$
MFC	$(3N + 1)$ (cost of each switch + cost of each diode) + N (cost of each capacitor)
Switch Mode	$4N$ (cost of each switch) + $2(N + 1)$ (cost of each diode + cost of each capacitor)
CCMMC	$(3N - 2)$ (cost of each switch + cost of each diode) + N (cost of each capacitor)
Figure 5.10	2 (cost of each switch) + $2N$ (cost of each diode + cost of each capacitor)

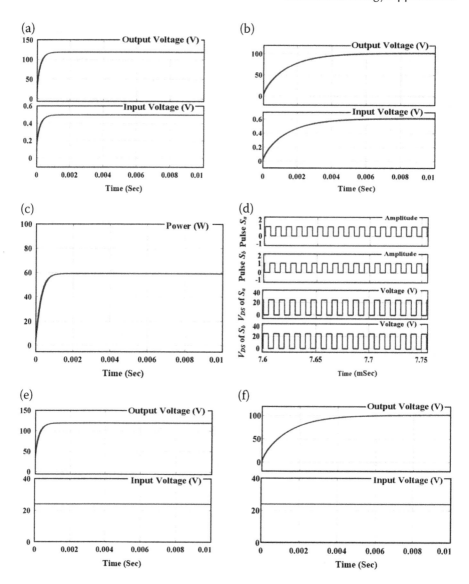

FIGURE 5.20 Simulation Results: (a) Current and Voltage Waveform without Considering Voltage Drop Across Devices, (b) Current and Voltage Waveform When Considering Voltage Drop Across Devices, (c) Power Waveform, (d) Voltage Across Switch, (e) Input and Output Voltage Waveform without Considering Voltage Drop Across Devices, (f) Input and Output Voltage Waveform When Considering Voltage Drop Across Devices.

considering diode and switch voltage drop, we find that a four-stage converter configuration provides a voltage conversion ratio of 5.

Thus, at the output terminal, 120 V achieved by using an input voltage of 24 V. When considering the 1-V voltage drop across diodes and switches, we find that

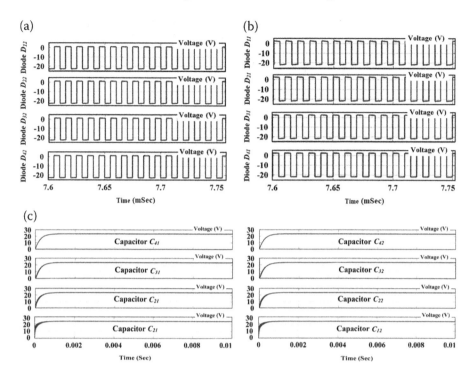

FIGURE 5.21   Simulation Results: (a)–(b) Voltage Across Diode Waveform, (c) Waveforms of Capacitors' Voltages.

TABLE 5.7
Hardware Components and Device Details

Component	Value
Control Switch	International Rectifier IRF250
Diode	BYQ28E
Load Resistance	60 W, 168 Ω
Capacitor	50 V, 220 uF
Driver IC	TLP250

100 V is achieved by using an input voltage of 24 V. We notice that when the switch is not conducting, the input voltage appears across the switch. In an ideal case, the voltage across all the capacitors is the same and equal to 24 V. The peak inverse voltage of the diode is equal to the input supply voltage, which is 24 V. Figure 5.21(a)–(b) depicts the voltage across a diode waveform. Figure 5.21(c) depicts the waveforms of the capacitors' voltages.

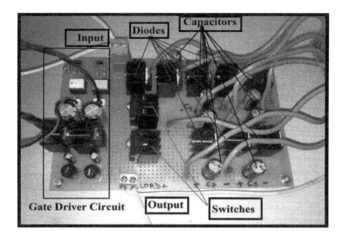

FIGURE 5.22   Hardware Implementation of Suggested DC-DC Multistage Converter.

The recommended converter is experimentally tested, and the numerical simulation results always show a good agreement with developed theoretical hypotheses. The components' used to build the converter and its peripheral control of the hardware setup given in Table 5.7. For switching the MOSFET, pulse-width

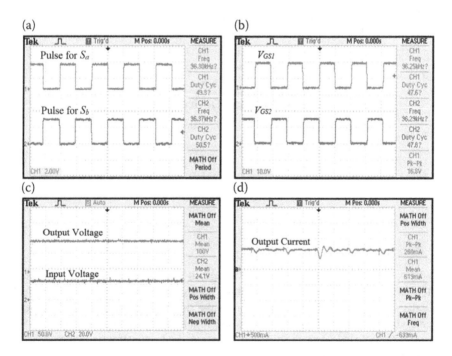

FIGURE 5.23   Experimental Results: (a) Switching Pulses Generated from PIC, (b) Pulses at Output of TLP-250, (c) Input and Output Voltage, (d) Waveform of Output Current.

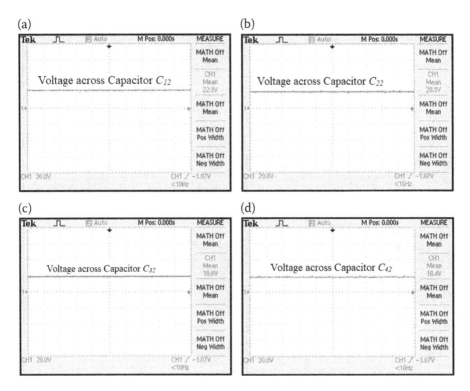

FIGURE 5.24 Experimental Results: (a) Voltage Across Capacitor C_{12}, (b) Voltage Across Capacitor C_{22}, (c) Voltage Across Capacitor C_{32}, (d) Voltage Across Capacitor C_{42}.

modulated (PWM) signals are obtained by using PIC18F45K20 digital processor. The output of the controller is assigned to driver TLP250, to generate the PWM signals for the MOSFET and elaborated by Figure 5.22.

Programmable Interface Controller (PIC) switching pulses are shown in Figure 5.23(a). The output pulses of driver TLP250 are shown in Figure 5.23(b). The input voltage and output voltage waveforms are shown in Figure 5.23(c), and we observe that 24 V is converted into a 100-V output. Figure 5.23(d) depicts the output current waveform. We notice that when compared to the input current, low current is flowing through the output terminal, and it is equal to 0.619 A.

Figure 5.24(a)–(d) depicts the waveforms of the voltage across capacitors C_{12}, C_{22}, C_{32}, and C_{42}. We observe that each capacitor voltage is the same and nearly equal to the input voltage of 24 V. Slightly less voltage is seen across the capacitors because of the voltage drop of diodes and switches. Figure 5.25(a)–(d) depicts the waveform of the voltage across diodes D_{12}, D_{22}, D_{32}, and D_{42}. We observe that the peak inverse voltage of each diode is the same and nearly equal to the input voltage of 24 V. We charge the capacitors of lower stages through the loop, which contains a smaller number of semiconductor devices. However, the

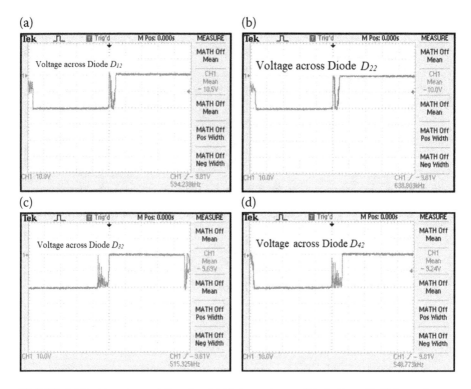

FIGURE 5.25 Experimental Results: (a) Voltage Across Diode D_{12}, (b) Voltage Across Diode D_{22}, (c) Voltage Across Diode D_{32}, (d) Voltage Across Diode D_{42}.

loop followed to charge other capacitors contains a number of semiconductor devices. Thus, a slight variation is noticed in the capacitor voltage.

To investigate the performance of the suggested converter, we operate the converter at a higher power (300 W), with an input voltage of 72 V and four stages ($K = 4$). Figure 5.26(a) depicts the current and voltage waveform without considering the diode and switch voltage drop. The received output voltage and current are 360 V and 0.84 A, respectively. It is notable that without considering the diode and switch voltage drop, the output voltage settles at 2 ms. Figure 5.26(b) depicts the power waveform, and it is seen that 300-W power is achieved at the output. Therefore, each stage contributes to a voltage of 72 V.

The suggested configuration can handle high power when utilizing a high-current/voltage source, higher rating switches, and a high number of stages. However, the system cost and converter increase with the rating of the converter. When considering the cost factor and volume of the converter (more stages are required for high power), the suggested converter is more suitable for low-power applications. The current rating of the MOSFET used and the source also limits the power of the designed prototype.

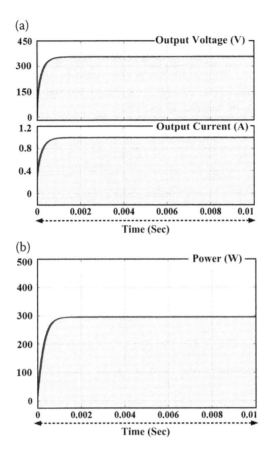

FIGURE 5.26 Simulation Results: (a) Current and Voltage Waveform without Considering the Voltage Drop Across Devices, (b) Current and Voltage Waveform When Considering the Voltage Drop Across Devices.

5.9 CONCLUSION

A self-balanced DC-DC multistage power converter configuration without magnetic components for photovoltaic applications was suggested and validated through hardware implementation. The recommended converter is suitable for photovoltaic applications, where the high voltage required augment without using inductor or transformer. The configuration also offers a workable solution in terms of modularity, control, and cost. The suggested configuration finds DC-link photovoltaic applications in DC-AC systems where voltage balancing of the capacitor is the major challenge. The configuration is also suitable for low-power applications since the transformer and inductor are not required. MPPT can quickly be implemented to improve the efficiency of the converter in terms of power extraction. The switch voltage is less and, thus, low-voltage rating switches can used for designing the configuration. The noticeable features of the suggested configuration are the absence of a magnetic component, the input current is continuous, the rating of semiconductor devices is low, the modular approach and levels can increase to enhance the voltage conversion ratio, and only two switches are required. The performance of the suggested converter configuration is validated from simulation and experimental results. The simulation and experimental results always showed good agreement with the theoretical analysis.

6 T-SC MPC: Transformer Switched-Capacitor-Based DC-DC Multistage Power Converter Configuration for Photovoltaic High-Voltage/Low-Current Applications

6.1 INTRODUCTION

In this chapter, for a transformer and switched-capacitor (T-SC)-based multistage power converter (MPC), we suggest a configuration for photovoltaic applications. The T-SC MPC configuration is an extension for the conventional boost converter. A SC and transformer are used to derive a T-SC converter configuration [10]. This T-SC converter configuration provides a good option for high-voltage/low-current photovoltaic applications. To achieve a high conversion ratio, we suggest a converter configuration combined with the feature of the conventional boost converter and T-SC cell. To convert maximum photovoltaic power into electricity, maximum power point tracking (MPPT) is compulsory [49–63] and used in the suggested system. The controlling mechanism when using the MPPT technique for a suggested T-SC MPC is also provided. The suggested converter provides a higher-voltage conversion ratio compared to the conventional DC-DC converter. The suggested T-SC MPC configuration's noticeable features are also discussed. In addition, steady-state analysis of the T-SC MPC configuration is provided and a configuration is compared with the newly addressed converter in terms of cost, ripples, efficiency, power range, voltage conversion ratio, and several components. Experimental and simulation results are provided to confirm the functionality, concept, and design of the suggested T-SC MPC configuration.

The conventional boost converter is not a workable solution for a photovoltaic system to achieve a high-voltage conversion ratio due to the large voltage across the control switch and leakage resistance of the inductor [236–238]. The conventional

---

$^{*}V_{out} = V_{o}$

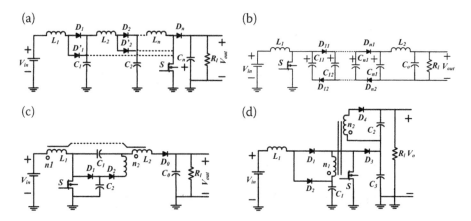

FIGURE 6.1 Power Circuits: (a) N-Stage Cascaded Boost Converter, (b) Voltage Multiplier-Based Extension for Boost Converter with Single Switch, (c) Step-Up Converter with Coupled Inductor, (d) Coupled-Inductor-Based Quadratic Boost Converter.

boost configuration is not suitable to boost the voltage over four because the characteristics and performance of the converter were deteriorating when the gate pulse with a large duty cycle is applied to the converter. A transformer and coupled inductor are other solutions to attain a high-voltage conversion ratio without using a large duty cycle to control the switches [239–241]. In the last decades, several DC boost converters were addressed by extending the conventional boost converter power circuit [71]. By using multiple stages of the conventional boost converter cascaded, we can design a boost converter to achieve a high-voltage conversion ratio [242,243]. Although high voltage is achieved by using a cascaded boost converter, to design this configuration, several capacitors, control switches, diodes, and inductors are required. The power circuit of the N-stage cascaded boost converter is shown in Figure 6.1(a) [243–245]. To overcome the drawback of the number of switches, a voltage multiplier-based extension for boost converter with the single switch shown in Figure 6.1(b)[244,245]. Similar to the cascaded boost converter, this configuration provides a high-voltage conversion ratio required by many capacitors and diode circuitry. A step-up converter with a coupled inductor to reduce the requirement of many capacitors and diode is discussed [246]. Figure 6.1(c) depicts the power circuit of the step-up converter with a coupled inductor [246]. Even if high voltage is achieved by using the transformer, the major drawbacks of this configuration are switching losses and high-voltage stress. A new coupled-inductor-based quadratic boost converter is shown in Figure 6.1(d)[163]. High voltage is achieved by using this configuration; the voltage across the switch is less.

6.2 TRANSFORMER AND SWITCHED-CAPACITOR (T-SC)-BASED MULTISTAGE POWER CONVERTER WITH MPPT

In this section, a T-SC is presented for low-current/high-voltage photovoltaic applications. Figure 6.2 depicts the T-SC MPC configuration with the functional

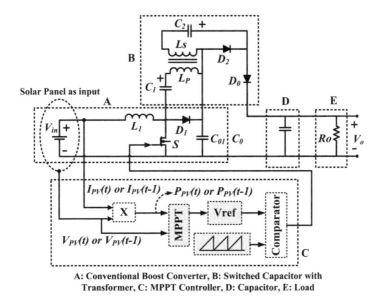

A: Conventional Boost Converter, B: Switched Capacitor with
Transformer, C: MPPT Controller, D: Capacitor, E: Load

FIGURE 6.2   T-SC MPC Configuration and Functional Block Diagram of Control Scheme
for Photovoltaic Applications.

block diagram of a control scheme for photovoltaic applications. The T-SC cell is
the extension for the conventional boost converter to achieve high voltage
at the output side. The T-SC cell combines the feature of the transformer and SC
cell. The T-SC MPC is a multistage power converter configuration, which combines
the feature of the T-SC cell and the conventional boost converter.

Figure 6.3 depicts the photovoltaic cell model in which a reverse diode is in
parallel with a current source, and a shunt and series resistance are connected. Shunt
resistance (R_{sh}) is added due to leakage current and series resistance and added
because of the barrier in the path of flow of charge from the n to p junction. The
current photovoltaic relation is provided in equation (6.1).

$$\left.\begin{array}{l} I = I_{ph} - I_d - I_{Sh},\ I_d = I_o \left(e^{\frac{V+IR_S}{nV_T}} - 1 \right),\ I_{Sh} = \frac{V+IR_S}{R_{Sh}} \\[2mm] V_T = \frac{kT}{q},\ V_T = 0.0259 Vat T = 25^\circ C \end{array}\right\} \tag{6.1}$$

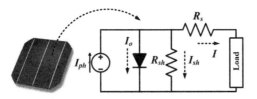

FIGURE 6.3   Photovoltaic Cell Model.

The current generated by a photovoltaic cell is I_{ph}, diode current is I_d, R_{sh} is shunt resistance, a current flowing through R_{sh} is I_{sh}, R_S is series resistance, the current flowing through R_S is I, photovoltaic cell terminal voltage is V, reverse saturation current is I_o, the ideal diode factor is n, thermal voltage is V_T, the Boltzmann constant is k, and the absolute temperature is T.

Within the cell, an electric field is created when solar radiation hits the plane surface of a photovoltaic cell or module. Current circulates by the charge generated because of the electric field and this current can used for the application. The value of the current depends on the solar radiation intensity and concentration. A high-value current is created if more electrons (here, called photons) can be permitted to travel free from the flat surface. Thus, high intensity and concentration of solar irradiance are required to generate more photons. Thus, it is compulsory to locate the MPP due to variation in solar irradiation and hotness.

Recently, it has been suggested that several MPPT systems can track the MPP by using various algorithms [58–61]. Each MPPT technique has its advantages and disadvantages. When the temperature and radiation changes, the choice of the algorithm changes based on the complexity, cost, accurate tracking, requirement of sensors, effectiveness in range, and tracking speed. To track the MPP, the most simple and popular algorithms are incremental conductance and perturb and observe (P&O) algorithms. The P&O algorithm is based on a hill-climbing idea; therefore, it is also called a P&O hill-climbing method or hill-climbing method. Because of low-cost, trouble-free implementation and simplicity, the P&O method is the most popular method to track the MPP. The method is based on the cyclic perturbation (decrease and increase) of the terminal voltage and last perturbation of power compared with present power. If the power rises, then perturbation moves in a similar direction; if it does not rise, then it will move in the reverse direction. In this technique, the sign of the previous perturbation and the sign of the previous increment in the power are used to choose what the next perturbation should be.

Figure 6.4(a)–(c) depicts the MPPT algorithm for the suggested scheme with I-V and P-V characteristics of photovoltaic applications.

To pull out the maximum power, the P&O MPPT-based control algorithm is used to trace the MPP for the presented T-SC MPC. The noticeable characteristics of the T-SC MPC configuration include the following:

- The voltage conversion ratio is high.
- The input current is continuous.
- It requires one switch to design the T-SC converter.
- One input source converter configuration
- Low-voltage rated switch
- Single transformer with untapped terminals and a single inductor

Figure 6.5(a) depicts the boost converter with the circuitry of a voltage doubler or SC. Diodes D_2 and D_0 and capacitors C_0 and C_1 are connections that form a voltage doubler circuit. Figure 6.5(b) depicts the power circuit of the T-SC MPC config-uration. The T-SC MPC is derived by utilizing the T-SC cell at the output port of the traditional boost converter. In the T-SC cell, the transformer is used in the

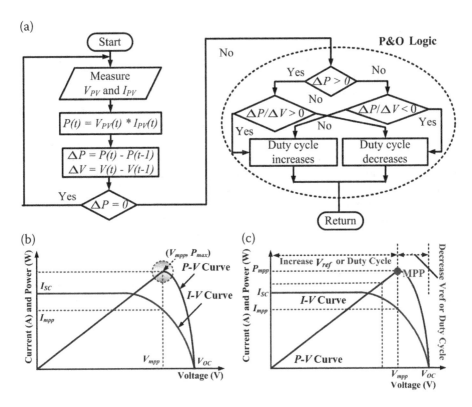

FIGURE 6.4 MPPT Technique for T-SC MPC: (a) MPPT Algorithm for Suggested Scheme, (b) *I-V* and *P-V* Photovoltaic Cell Characteristics, (c) Idea to Follow MPP to Pull Out Maximum Power from Photovoltaic Module or Array.

middle of the SC to attain a high-voltage conversion ratio. In Figure 6.5(b), diode D_1, capacitor C_{01}, inductor L_1, and switch S are elements of a traditional boost converter. Conventional boost converter circuitry is directly fed from the input

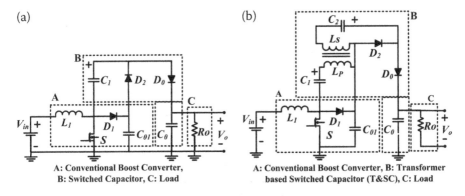

FIGURE 6.5 (a) Boost Converter with Circuitry of Voltage Doubler or SC, (b) Power Circuit of T-SC MPC Configuration.

photovoltaic supply. A transformer, diodes D_0 and D_2, and capacitors C_1 and C_2 are the elements of a T-SC cell.

The transformer primary winding (L_P) has one leg linked via the C_1 capacitor at the inductor of the traditional boost converter and another leg directly linked at the cathode of the D_1 diode. The transformer secondary winding (L_S) has one leg directly linked to the anode of the D_2 diode and another leg connected to the C_2 capacitor. The workings of the T-SC MPC are split into two dominant modes: first, when switch S is conducting, and second, when switch S is not conducting. To discuss the continuous conduction mode (CCM) of the T-SC MPC configuration, the two main modes are split into five sub-modes. Figure 6.6 depicts the characteristic waveforms of the T-SC MPC configuration in the CCM mode.

In the first sub-mode (time t_a to t_b), control switch S is not conducting and the demagnetization of the inductor takes place. Diode D_1 and capacitor C_{O1} are charged by the energy stored in inductor L_1. Through diode D_0, the energy of capacitors C_2 and C_1 are transferred to the output capacitor C_0. For this mode, the converter equivalent circuit is depicted in Figure 6.7(a).

In this sub-mode, the negative voltage is present across D_2, and the positive voltage is present across diodes D_0 and D_1. Hence, diode D_2 is reverse-biased and diodes D_0 and D_1 are forward-biased. The equation for the first sub-mode is given in equation (6.2).

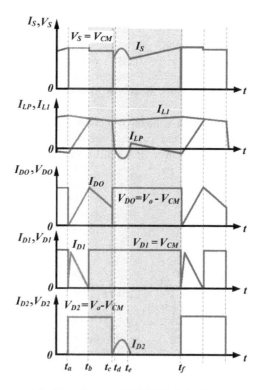

FIGURE 6.6 Characteristic Waveforms of T-SC MPC Configuration in CCM Mode.

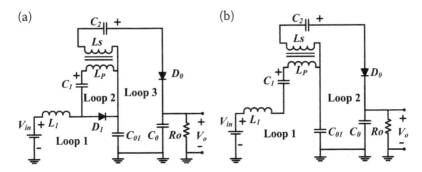

FIGURE 6.7 T-SC MPC Equivalent Circuit When Switch S Is Not Conducting: (a) First Sub-mode (Time t_a to t_b), (b) Second Sub-mode (Time t_b to t_c).

$$\left. \begin{aligned} V_{in} - VC_{01} - V_{L1} = 0 &\rightarrow Loop - 1 \\ V_{in} + VC_1 - V_{L1} - VC_{01} - V_{LP} = 0 &\rightarrow Loop - 2 \\ V_{LS} + VC_{01} - V_{CO} + VC_2 = 0 &\rightarrow Loop - 3 \end{aligned} \right\} \quad (6.2)$$

In the second sub-mode (time t_b to t_c), switch S is not conducting and demagnetization of the inductor still takes place. Through diode D_1, the energy of the inductor L_1 is not directly fed to the capacitor C_{01}, but is fed through the transforming winding. Through diode D_O, the stored energy of capacitors C_2 and C_1 and transformer are fed to charge the output C_O capacitor. For this mode, the converter equivalent circuit is depicted in Figure 6.7(b). In this sub-mode, the negative voltage is present across diodes D_1 and D_2 and a positive voltage is present across diode D_O. Hence, diodes D_1 and D_2 are reversed-biased and diode D_O is forward-biased. The equation for the second sub-mode is given in equation (6.3).

$$\left. \begin{aligned} V_{in} + VC_1 - V_{L1} = VC_{01} + V_{LP} &\rightarrow Loop - 1 \\ VC_2 + V_{LS} = -VC_{01} + V_{CO} &\rightarrow Loop - 2 \end{aligned} \right\} \quad (6.3)$$

In the third sub-mode (time t_c to t_d), switch S is conducting and the magnetization of inductor L_1 takes place through switch S. Through diode D_O, the energy of transformer windings, and capacitor C_2, C_1 is fed to charge output C_O capacitor. However, the current that starts at diode D_O is decreasing and the transformer restricts the slope of the current (d_i/d_t). Therefore, the reverse recovery problem of diode current is reduced. For this mode, the converter equivalent circuit is depicted in Figure 6.8(a). In this sub-mode, negative voltage is present across diodes D_1 and D_2 and positive voltage is present across diode D_O. Hence, diodes D_1 and D_2 are reversed-biased, and diode D_O is forward-biased. The equation for the third sub-mode is given in equation (6.4).

$$\left. \begin{aligned} -V_{L1} + V_{in} = 0 &\rightarrow Loop - 1 \\ VC_1 - VC_{01} - V_{LP} = 0 &\rightarrow Loop - 2 \\ V_{LS} + VC_{01} + VC_2 = V_{CO} &\rightarrow Loop - 3 \end{aligned} \right\} \quad (6.4)$$

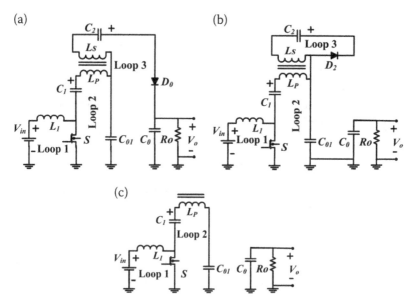

FIGURE 6.8 T-SC MPC Converter Equivalent Circuit When Switch S Is Conducting:
(a) Third Sub-mode (Time t_c to t_d), (b) Fourth Sub-mode (Time t_d to t_e), (c) Fifth Sub-mode
(Time t_e to t_f).

In the fourth sub-mode (time t_d to t_e), switch S is conducting and the magnetization
of inductor L_1 continues through switch S. Through diode D_2, the L_S transformer
winding energy is transferred to charge the C_2 capacitor. The leakage inductance
of the transformer limits the current, and conversion of energy takes place in
a resonant way. Finally, at the end of this sub-mode, capacitor C_2 is fully
charged and diode D_2 is not conducting. For this mode, the converter equivalent
circuit is depicted in Figure 6.8(b). In this sub-mode, the negative voltage is present
across diodes D_0 and D_1 and a positive voltage is present across diode D_2. Hence,
diodes D_0 and D_1 are reversed-biased and diode D_2 is forward-biased. The equation
for the fourth sub-mode is given in equation (6.5).

$$\left.\begin{array}{l} - V_{L1} + V_{in} = 0 \rightarrow Loop - 1 \\ VC_1 - VC_{01} - V_{LP} = 0 \rightarrow Loop - 2 \\ VC_2 = -V_{LS} \rightarrow Loop - 3 \end{array}\right\} \quad (6.5)$$

In the fifth sub-mode (time t_e to t_f), switch S is conducting and the magnetization of
inductor L_1 continues through switch S. In this sub-mode, diode D_2 is reverse-
biased; thus, the L_S transformer winding energy is not transferred to charge the C_2
capacitor. The C_0 capacitor transfers its energy to load. For this mode, the converter
equivalent circuit is depicted in Figure 6.8(c). In this sub-mode, the negative voltage is present across diodes D_0, D_1, and D_2. Hence, diodes D_0, D_1, and D_2 are
reversed-biased, and no diode is forward-biased. The equation for the fifth sub-
mode is given in equation (6.6).

$$\left.\begin{array}{l} V_{in} - V_{L1} = 0 \rightarrow Loop - 1 \\ VC_1 = V_{LP} + VC_{01} \rightarrow Loop - 2 \end{array}\right\}$$ (6.6)

Equation (6.7) is used to calculate the voltage across the switch and voltage conversion ratio of a boost converter with a voltage doubler, shown in Figure 6.4(a), where one switching cycle time is T. The graph of the voltage conversion ratio versus duty cycle for a boost converter with voltage doubler is shown in Figure 6.9(a).

$$\left.\frac{V_o}{V_{in}} = 2\frac{1}{1-D} = 2\frac{1}{1 - \frac{T_{on}}{T}}, \ V_{DS} = V_{in}\frac{1}{1-D} = \frac{1}{2}V_o = V_{in}\frac{1}{1 - \frac{T_{on}}{T}}\right\}$$ (6.7)

The graph of the ratio of the drain-to-source voltage and input voltage versus duty cycle for a boost converter with voltage doubler is also shown in Figure 6.9(a). We find that the voltage across the switch is exactly half of the output terminal voltage.

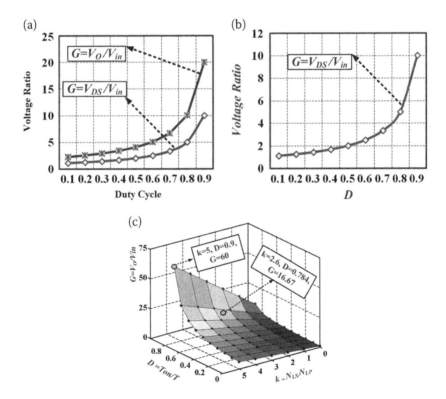

FIGURE 6.9 Plots: (a) Graph of Voltage Conversion Ratio Versus Duty Cycle for Boost Converter with Voltage Doubler Configuration, (b) Graph of Ratio of Drain to Source Voltage and Input Voltage Versus Duty Cycle for T-SC MPC Configuration, (c) Voltage Conversion Ratio and Turns Ratio Versus Duty Cycle for T-SC MPC Configuration.

For example, at $D = 0.6$, the voltage conversion ratio is 2.5 and $V_{DS}/V_{in} = 1.25$. Therefore, to avoid the problem of the switch, the drain-to-source voltage rating must be more than 1.25 times the input voltage when the switch is operated at 60% duty cycle.

Equation (6.8) is used to calculate the voltage across the switch and the voltage conversion ratio of the T-SC MPC configuration, where one switching cycle time is T. The graph of the ratio of the drain-to-source voltage and input voltage versus duty cycle for the T-SC MPC configuration is shown in Figure 6.9(b). The graph of the voltage conversion ratio and turns ratio versus duty cycle for the T-SC MPC configuration is shown in Figure 6.9(c). We notice that the slope of the voltage conversion ratio linearly increases with the turns ratio and duty cycle. First, we see that the 16.67 voltage conversion ratio is achieved when $k = 2.6$, the duty cycle is 78.4%, and a 60% voltage conversion ratio is achieved when $k = 5$ and the duty cycle is 90%. Second, we find that the required rating of the switch to design the T-SC MPC is precisely the same as the switch rating of a boost converter with a voltage doubler.

$$\left. \begin{aligned} \frac{V_o}{V_{in}} &= (k + 1)\frac{1}{1 - D} = (k + 1)\frac{1}{T - \frac{T_{on}}{T}} \\ V_{DS} &= V_{in}\frac{1}{1 - D} = V_o\frac{1}{1 + k} = V_{in}\frac{1}{T - \frac{T_{on}}{T}} \\ k &= \frac{transformer\ Secondary\ winding\ Turns,\ N_{LS}}{transformer\ Primary\ winding\ Turns,\ N_{LP}} \end{aligned} \right\} \qquad (6.8)$$

6.3 ANALYSIS OF THE T-SC MPC CONFIGURATION

In this section, the T-SC MPC configuration analysis in steady-state mode is explained and the mathematical equation is derived from calculating the voltage conversion ratio, efficiency, and conversion losses. The steady-state mode with the following assumptions is considered to analyze the T-SC MPC configuration:

- Pure DC source at the input side
- V_{D1} is the voltage drop across diode D_1.
- R_{D1} is theon-state resistance of diode D_1.
- If V_{D1} and R_{D1} are equal to 0, then the efficiency of the diode is 0.
- The control switch ON-state resistance is R_S.
- The internal resistance of the inductor L_1 is R_{L1}.
- The switching frequency is f_s.
- Minimal ripple across the capacitor
- Copper loss, diode conduction loss, and iron loss are included in the T-SC cell loss.
- T-SC cell equivalent resistance is R_{T-SC}

The equivalent steady-state circuit of the T-SC MPC when the switch is conducting (turned ON) and not conducting (turned OFF) is shown in Figure 6.10(a)–(b), respectively. Consider i_s, i_{L1}, and ic_{01} are currents flowing through switch S, L_1 inductor,

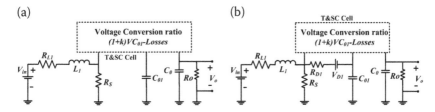

FIGURE 6.10 Power Circuits: (a) Equivalent Steady-State Circuit of T-SC MPC When Switch Is Conducting (Turned ON), (b) Equivalent Steady-State Circuit of T-SC MPC When Switch Is Not Conducting (Turned OFF).

and C_{01} capacitor, respectively. I_s, I_{L1}, and I_{C01} are the average current flowing through switch S, L_1 inductor, and C_{01} capacitor, respectively. The D_1 diode is reversed-biased and the inductor is magnetized by the input voltage (V_{in}) at the time switch S is conducting. The steady-state voltage and current are given as below

$$
\left.
\begin{aligned}
V_{in} - i_s R_s - i_{in} R_{L1} - v_{L1} &= 0,\ i_s = i_{in} \\
v_{L1} = -i_{in}(R_{L1} + R_s) + V_{in} &\approx -I_{in}(R_{L1} + R_s) + V_{in} \\
i_c(t) &= -V_{c01}(R_{T-SC})^{-1}
\end{aligned}
\right\}
\qquad (6.9)
$$

The D_1 diode is forward-biased and the inductor is demagnetized to charge the output C_{01} capacitor at the time switch S is not conducting, the characteristic equation is given by

$$
\left.
\begin{aligned}
-v_{L1} + V_{in} - V_{D1} - i_{in} R_{L1} - V_{c01} - i_{D1} R_{D1} &= 0,\ i_{D1} = i_{in} \\
v_{L1} = -V_{D1} + V_{in} - i_{in}(R_{D1} + R_{L1}) - V_{c01} &\approx -V_{D1} + V_{in} - I_{in}(R_{D1} + R_{L1}) - V_{c01} \\
i_c(t) &= i_{in} - V_{c01}(R_{T\&SC})^{-1}
\end{aligned}
\right\}
$$
$$(6.10)$$

To calculate the equation of the voltage conversion ratio, capacitor charge, and inductor volt, a second balance method is applied.

$$
\left.
\begin{aligned}
V_{c01} &= (-(1-D)V_{D1} + V_{in})\left(\frac{1}{1-D}\right) \times \left(\frac{R_{T-SC}(1-D)^2}{R_{L1} + (1-D)^2 R_{T-SC} + (1-D)R_{D1} + DR_s}\right) \\
\frac{V_{c01}}{V_{in}} &= (1 - \frac{V_{D1}}{V_{in}}(1-D))\left(\frac{1}{1-D}\right) \times \left(\frac{1}{\frac{DR_s + R_{L1} + R_{D1}(1-D)}{R_{T-SC}(1-D)^2} + 1}\right)
\end{aligned}
\right\}
\qquad (6.11)
$$

$$
V_o = -losses \text{ of transformer} + (1+k)V_{c01}
\qquad (6.12)
$$

$$V_o = \left(\frac{1}{1-D}\right)(-(1-D)V_{D1} + V_{in})(1+k) \times \left(\frac{(1-D)^2 R_{T-SC}}{R_{L1} + (1-D)^2 R_{T-SC} + (1-D)R_{D1} + DR_s}\right)$$

$$\frac{V_o}{V_{in}} = \left(\frac{1}{1-D}\right)(1+k)\left(1 - \frac{V_{D1}}{V_{in}}(1-D)\right) \times \left(\frac{1}{1 + \frac{DR_s + R_{L1} + R_{D1}(1-D)}{R_{T-SC}(1-D)^2}}\right) \tag{6.13}$$

$$Conversion\ Losses = \left(\frac{k+1}{1-D}\right) \times \frac{V_{D1}(1-D)}{V_{in}}\left(\frac{1}{\frac{DR_s + R_{L1} + R_{D1}(1-D)}{R_{T-SC}(1-D)^2} + 1}\right) \tag{6.14}$$

$$\eta,\ efficiency = \frac{(1 - \frac{V_{D1}(-D+1)}{V_{in}})}{\frac{DR_s + R_{L1} + R_{D1}(1-D)}{R_{T-SC}(1-D)^2} + 1} \tag{6.15}$$

When the efficiency of the components 100% (components are ideal) then,

$$Losses = 0\%,\ \eta\,(efficiency) = 100\%$$
$$then\frac{V_o}{V_{in}} = \left(\frac{1}{1-D}\right)(k+1) \tag{6.16}$$

When ideal components are considered, equation (6.16) provides the voltage conversion ratio of the T-SC MPC configuration.

6.4 COMPARISON OF THE T-SC MPC CONFIGURATION WITH INDUCTOR-LESS AND TRANSFORMER-LESS DC-DC CONVERTERS

In this section, the T-SC MPC configuration is compared with inductor-less and transformer-less DC-DC converters. In Table 6.1, T-SC converters are compared with each other in terms of number of inductors, number of inductors, and the voltage conversion ratio.

We find that the voltage conversion ratio of the n-stage cascaded boost converter depends on the duty cycle and the number of stages. The number of diodes, $2n-1$, single switch, n number of capacitors, and n number of inductors are required to design the n-stage cascaded boost converter. We note that the voltage conversion ratio of the multiplier-based extension for the boost converter with 1 switch depends on the duty cycle and the even and odd numbers of stages. The $2n$ number of diodes, 1 switch, $2n+1$ number of capacitors, and 2 inductors are required to design the multiplier-based extension for the boost converter. We find that the voltage conversion ratio of the step-up converter with a coupled inductor depends on the duty cycle and coupled inductor coupling coefficient (k). Two diodes,

TABLE 6.1

Comparison of T-SC Converter with DC-DC Converters

Converter	Voltage Conversion Ratio (V_O/V_{in})	L	C	D
Figure 6.1(a)	$\frac{1}{(1-\Delta)^n}$	n	n	$2n-1$
Figure 6.1(b)	$\frac{n+\Delta}{1-\Delta}$, $n = 1, 3 \ldots$ or $\frac{n+1+\Delta}{1-\Delta}$, $n = 2, 4 \ldots$	2	$2n+1$	$2n$
Figure 6.1(c)	$\frac{1+(1+k)\Delta}{1-\Delta}$, $k = \frac{n_2}{n_1}$	3 (2 inductor are coupled)	3	3
Figure 6.1(d)	$\frac{1+k\Delta}{(1-\Delta)^2}$, $k = \frac{n_2}{n_1}$	3 (2 inductors are coupled)	3	4
T-SC MPC Configuration	$\frac{1+k}{1-\Delta}$, $k = \frac{n_{LS}}{n_{LP}}$	1 inductor with 1 transformer	4	3

1 switch, 3 capacitors, and 3 inductors (2 coupled and 1 without coupling) are required to design a step-up converter with the coupled inductor.

We observe that the voltage conversion ratio of the coupled inductor-based quadratic boost converter depends on the duty cycle and the coupled inductor coupling coefficient (k). Four diodes, 1 switch, 3 capacitors, and 3 inductors (2 coupled and 1 without coupling) are required to design a coupled inductor-based quadratic boost converter. It is also investigated that 3 diodes, 1 inductor, 1 switch, and 3 capacitors are required to design a boost converter with a voltage doubler. The T-SC MPC voltage conversion ratio depends on the duty cycle and turns ratio of the transformer (k). Three diodes, 1 switch, 1 inductor, 4 capacitors, and 1 transformer are required to design the T-SC MPC configuration. Compared with recent converters, the T-SC MPC configuration has a higher voltage conversion ratio and requires a smaller number of switches.

In Table 6.2, the T-SC MPC is compared with DC-DC converters in terms of cost, where C_C, C_L, C_D, C_S, and C_{CL} are the cost of 1 capacitor, 1 inductor, 1 diode, 1 switch, and 1 coupled inductor, respectively. Notice that the cost is less to design

TABLE 6.2

Cost of the DC-DC Converter

DC-DC Converter	Cost of Converter
Figure 6.1(a)	$n(C_L+C_C)+(2n-1)C_D+C_S$
Figure 6.1(b)	$2C_L+(2n+1)C_C+(2n)C_D+C_S$
Figure 6.1(c)	$C_{CL}+C_L+3(C_C+C_D)+C_S$
Figure 6.1(d)	$C_{CL}+C_L+3C_C+4C_D+C_S$
T-SC MPC Configuration	$C_{CL}+C_T+4C_C+3C_D$

the T-SC converter. The graph of the voltage conversion ratio versus duty cycle for an n-stage cascaded boost converter and multiplier-based extension for boost converter (for stages 1 to 5) is depicted in Figure 6.11(a)–(b). We observe that if the number of stages of the n-stage cascaded boost converter and multiplier-based

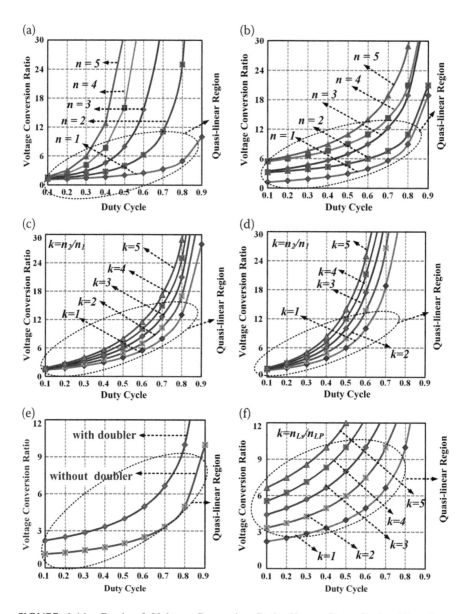

FIGURE 6.11 Graph of Voltage Conversion Ratio Versus Duty Cycle: (a) n-Stage Cascaded Boost Converter, (b) Multiplier-Based Extension for Boost Converter, (c) Step-Up Converter with Coupled Inductor, (d) Coupled Inductor-Based Quadratic Boost Converter, (e) Boost Converter with Voltage Doubler, (f) T-SC MPC.

extension for boost converter is increased, then the quasi-linear region decreases and the voltage conversion ratio is increases. The graph of the voltage conversion ratio versus duty cycle for a step-up converter with coupled inductor and coupled inductor-based quadratic boost converter (with coupling coefficient, $k = 1$ to 5) is depicted in Figure 6.11(c)–(d). Notice that if the coupling coefficient (k) coupled inductor increases, then the quasi-linear region decreases and the voltage conversion ratio sharply increases.

The graph of the voltage conversion ratio versus duty cycle for boost converter with voltage doubler is depicted in Figure 6.11(e). We find that if the doubler attaches to the conventional boost converter, then the double voltage conversion ratio is achieved and the quasi-linear region decreases when compared to the conventional boost converter. Figure 6.11(f) depicts the graph of the voltage conversion ratio of the T-SC MPC configuration with transformer turns ratio $k = 1$ to 5. We find that if the transformer turns ratio increases, then the quasi-linear region decreases and the voltage conversion ratio is sharply increases.

Figure 6.12(a) depicts the graph of the ratio of voltage across the capacitor to input supply voltage versus duty cycle for n-stage cascaded boost converter with considering the number of stages, $n = 1$ to 5. We observe that if the duty cycle and the number of stages increases, then the required rating of the capacitor increases. Therefore, the rating of the capacitor of the n-stage cascaded boost converter is higher than the rating of the capacitor of the $n-1$ stage cascaded boost converter. Figure 6.12(b) depicts the graph of the ratio of voltage across the capacitor to input supply voltage versus duty cycle for the multiplier-based extension for boost converter when considering the number of stages, $n = 1$ to 5. Notice that if the duty cycle and number of multiplier stages increases, then the required rating of the capacitor increases. Figure 6.12(c) depicts the graph of the ratio of voltage across the capacitor to input supply voltage versus duty cycle for a step-up converter with coupled inductor by considering the coupling factor, $k = 1$ to 5. We find that voltage across the capacitor depends on the duty cycle but is independent of the coupling factor.

Figure 6.12(d) depicts the graph of the ratio of voltage across the capacitor C_{02} (VC_{02}) to the voltage across the capacitor C_{01} (VC_{01}) versus duty cycle for the coupled inductor-based quadratic boost converter by considering the coupling factor, $k = 1$ to 5.

The voltage across the capacitor depends on the duty cycle as well as the coupling factor. Thus, if the duty cycle and the coupling coefficient increases, then the required rating of the capacitor increases. Figure 6.12(e) depicts the graph of the ratio of voltage across capacitor C_{01} to input supply voltage versus duty cycle for the T-SC MPC by considering the transformer turns ratio, $k = 1$ to 5. The voltage across the capacitor depends on the duty cycle but is independent of the transformer turns ratio. Figure 6.12(f) shows the comparison of the T-SC MPC with recent inductor-less and transformer-less converters (discussed in the introduction section of this chapter) in terms of the voltage conversion ratio. The T-SC MPC provides a higher voltage at the output terminal compared to inductor-less and transformer-less converters. The voltage across the switch of the T-SC MPC is lower compared to

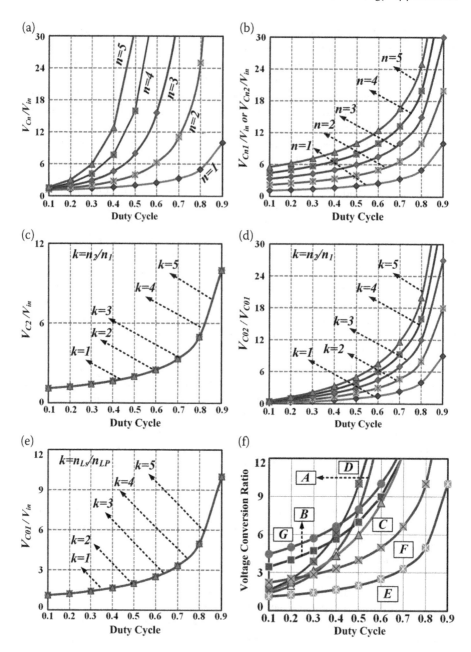

FIGURE 6.12 Graph of Voltage Conversion Ratio Versus Duty Cycle: (a) n-Stage Cascaded Boost Converter [A], (b) Multiplier-Based Extension for Boost Converter [B], (c) Step-Up Converter with Coupled Inductor [C], (d) Coupled Inductor-Based Quadratic Boost Converter [D], (e) T-SC MPC [G], (f) Comparison of T-SC MPC and Converters, [E] Conventional Boost Converter, [f] Conventional Boost Converter with Voltage Doubler.

the voltage across the switch of inductor-less and transformer-less converters. Thus, a low-rated switched is required for the T-SC MPC configuration.

6.5 VALIDATION OF THE T-SC MPC CONFIGURATION

The functionality and design of the T-SC MPC configuration is tested and verified by simulation software Matrix Laboratory 9.0 (R2016) and Table 6.4 tabulates the designed parameter. The component of the T-SC MPC configuration is designed and selected based on the (6.17)–(6.24).

$$Duty\ cycle,\ \Delta = \frac{-V_{in}(k+1) + V_o}{V_o} = 78.4\%\ or\ 0.784 \tag{6.17}$$

$$Switch\ Voltage,\ V_{DS} = \frac{V_{in}}{1 - T_{ON}f} = 69.44V \approx 70V \tag{6.18}$$

$$Diode\ D_0\ Voltage,\ V_{D0} = V_{in}\frac{1}{1 - T_{ON}f} \times \frac{n_{LS}}{n_{LP}} = \frac{k}{1 - fT_{ON}}V_{in} = 180.5V \approx 181V \Big\} \tag{6.19}$$

$$V_{D1} = \frac{V_{in}}{1 - T_{ON}f} = \frac{15}{1 - 0.784} = 69.44V \approx 70V \tag{6.20}$$

$$\left. \begin{aligned} Diode\ D_2\ Voltage,\ V_{D2} &= \frac{V_{in}}{1 - T_{ON}f} \times \frac{n_{LS}}{n_{LP}} \\ &= \frac{V_{in}}{1 - T_{ON}f}k = 180.5V \approx 181V \end{aligned} \right\} \tag{6.21}$$

$$Inductor,\ L_1 = V_{in}\frac{T_{ON}f}{f_1 \times \Delta iL} = 117.6uH \tag{6.22}$$

$$L_P = \frac{T_{ON}f}{f \times \Delta iL_1}V_{in} = \frac{15 \times 0.784}{5 \times 20000} = 117.6uH \tag{6.23}$$

$$L_S = V_{in} \times k^2 \times \frac{T_{ON}f}{f \times \Delta iL_1} = 794.97uH \tag{6.24}$$

The value of the capacitor to design the T-SC MPC configuration is deduced by using switching frequency, voltage ripple across the capacitor, capacitor current, and transformer turns ratio. The input voltage waveform and the output voltage waveform are depicted in Figure 6.13(a). We observe that 250-V, constant DC is achieved with the 15-V input voltage. Therefore, a 16.67 voltage conversion ratio is achieved.

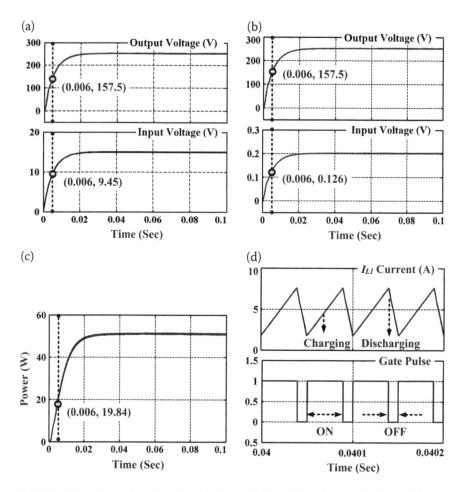

FIGURE 6.13 Simulation Results: (a) Input Voltage Waveform and Output Voltage Waveform, (b) Output Current and Output Voltage Waveform, (c) Output Power Waveform, (d) Current Flowing through the L_1 Inductor.

In transient response, the time constant (τ) for input voltage and output voltage waveform is 0.006 seconds. There is a 157.5 output voltage, and the 9.45 input voltage is observed at a given time constant. Approximately 0.035 seconds are required to attain a 250-V constant DC. The output current and output voltage waveform are shown in Figure 6.13(b). In transient response, the time constant (τ) for output current and output voltage waveform is 0.006 seconds. A 157.5 output voltage and 0.126 output current are observed at a given time constant (τ). Approximately 0.035 seconds are required to attain a 0.2-A constant current.

The output power waveform is shown in Figure 6.13(c) and investigated that 19.84-W power is observed at 0.0006 seconds. Approximately 0.035 seconds are required to attain 50-W constant power. The current is flowing through the L_1 inductor shown in

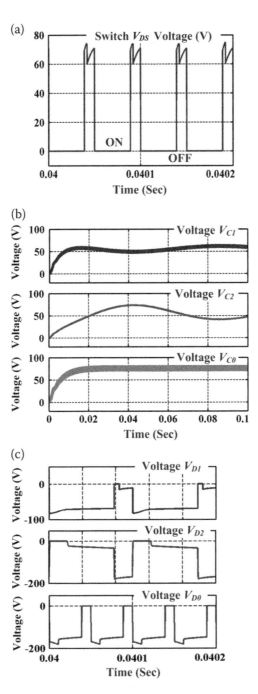

FIGURE 6.14 Simulation Results: (a) Voltage Waveform Across Switch, (b) C_{01}, C_1, and C_2 Capacitors' Voltage, (c) D_0, D_2, and D_1 Diodes' Voltage.

Figure 6.13(d) with gate pulse. The slope of the inductor current waveform is measured by (6.17). We find that positive slope is achieved when the switch is conducting and negative slope achieved when the switch is not conducting. The voltage waveform across the switch is shown in Figure 6.14(a). We find that 70 V has appeared across the switch when it is not conducting. In the switch voltage waveform, a small fluctuation is observed because the voltage across diodes is varied in sub-mode.

Figure 6.14(b) depicts the voltage waveform of the capacitors C_2, C_1, and C_{01}. Approximately 70 V is observed across the C_{01} capacitor. The ripple in the C_1, C_2, and C_{01} capacitors' voltage is 1.2, 1.31, and 2.1 V (in Figure 6.14(b) ripples look high due to use of the broad line for high quality). Figure 6.14(c) shows the voltage waveform across diodes D_1, D_0, and D_2 and we find that when the switch is not conducting the diode, D_0 is forward-biased. Higher voltage is observed across diodes D_2 and D_0 compared to the voltage across diode D_1. The hardware implementation of the T-SC MPC configuration is done and the details of the components with parameters are provided in Table 6.3. The hardware prototype of the T-SC MPC is shown in Figure 6.15(a). Figure 6.15(b)–(c) shows the output voltage and input voltage waveforms. We find that 249.6-V output is achieved from the 15.1-V input.

$$\left.\begin{array}{l} \dfrac{di_{L1}}{dt} = Positive\ Slope = VL_{1ON} \times \dfrac{1}{L_1} \\[2mm] \dfrac{di_{L1}}{dt} = Negative\ Slope = VL_{1OFF} \times \dfrac{1}{L_1} \end{array}\right\} \qquad (6.25)$$

6.6 THE T-SC MPC CONFIGURATION WITH A VOLTAGE MULTIPLIER

The voltage multiplier is combined with the T-SC MPC configuration to achieve a very high voltage at the output terminal. Voltage multipliers provide a good option with the T-SC converter to achieve a high-voltage conversion ratio. Figure 6.16 depicts the power circuit of the T-SC MPC with the multiplier circuit.

TABLE 6.3

Experimental and Simulation Parameters

Parameter	Value
Input voltage (V_{in}), output voltage (V_o)	15 V, 250 V
Input current (I_{in}), output current (I_o)	3.34 A, 0.2 A
Power (P)	50 W
Duty cycle (D), diode	0.784, MUR 860
Switching frequency (f_s), switch	20 kHz, IRF 540
Switch drain-to-source voltage (V_{DS})	70 V (minimum voltage)
Inductor (L_1)	117.6 uh
Transformer windings	LP = 117.6 uH, LS = 794.97 uH turns ratio, k = 2.6
Capacitors (C_1, C_2)	2.5 uF, 250 V
Capacitors (C_0, C_{01})	100 uF, 400 V

(a)

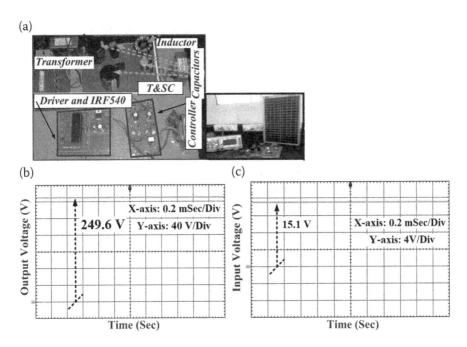

(b)

Output Voltage (V)

X-axis: 0.2 mSec/Div
Y-axis: 40 V/Div

249.6 V

Time (Sec)

(c)

Input Voltage (V)

15.1 V

X-axis: 0.2 mSec/Div
Y-axis: 4V/Div

Time (Sec)

FIGURE 6.15 Experimental Work: (a) Hardware Setup, (b) Output Voltage, (c) Input Voltage.

The mathematical equation to calculate the voltage conversion ratio is given in equation (6.26).

$$
\left.
\begin{array}{l}
\dfrac{V_{C01}}{V_{in}} = (1+k)\dfrac{1}{1-\frac{T_{on}}{T}},\ \dfrac{V_{C03}}{V_{in}} = (1+k)\dfrac{2}{1-\frac{T_{on}}{T}} \\[3mm]
\dfrac{V_{C05}}{V_{in}} = (1+k)\dfrac{3}{1-\frac{T_{on}}{T}},\ \dfrac{V_{C02N-1}}{V_{in}} = \dfrac{V_o}{V_{in}} = (1+k)\dfrac{N}{1-\frac{T_{on}}{T}} \\[3mm]
V_{DS} = \dfrac{1}{1+k}V_{Co1} = \dfrac{N^{-1}V_o}{(1+k)} = \dfrac{1}{1-\frac{T_{on}}{T}}V_{in},\ k = \dfrac{Secondary\ winding\ Turns\ (N_{LS})}{primary\ winding\ Turns\ (N_{LP})} \\[3mm]
N = number\ of\ voltage\ multiplier\ stage
\end{array}
\right\}
\quad (6.26)
$$

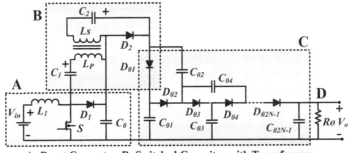

**A: Boost Converter, B: Switched Capacitor with Transformer
C: Voltage Multiplier, D: Load**

FIGURE 6.16 Power Circuit of T-SC MPC Configuration with Voltage Multiplier.

We find that the drain-to-source voltage of the switch is not affected due to the addition of the voltage multiplier. Thus, the voltage multiplier circuitry can attach to the output side of the T-SC MPC configuration without disconnecting the main circuit components to increase the voltage conversion ratio with have less voltage stress across the switch.

6.7 CONCLUSION

The T-SC MPC was discussed for high-voltage and low-current photovoltaic applications. A transformer-based SC is designed to achieve a high-voltage conversion ratio. This converter combines the features of the traditional boost converter, voltage multiplier, SC, and transformer to achieve a high-voltage conversion ratio. A noticeable qualities of the T-SC MPC are a high-voltage conversion ratio and continuous input current. Only one switch and source are needed, the voltage across the switch is low, and only one inductor and one transformer are required. When the T-SC configuration is compared with other converters, a T-SC MPC configuration requires a smaller number of switches.

Furthermore, the cost of the T-SC MPC configuration is less and also easily modified by using a voltage multiplier to achieve high voltage. The concept and working of the T-SC configuration is verified by simulation and experimental investigation.

7 New Cockcroft Walton Voltage Multiplier-Based Multistage/Multilevel Power Converter Configuration for Photovoltaic Applications

7.1 INTRODUCTION

In this chapter, a new Cockcroft Walton (CW) voltage multiplier-based multistage/ multilevel power converter (CW-VM-MPC) configuration is suggested for photovoltaic applications. We suggest the interleaved converter configuration of a Nx multilevel boost converter, also called a Nx IMBC. Therefore, this suggested configuration combines the features of the boost converter with interleaved structure and a CW voltage multiplier. In most critical photovoltaic cases, voltage produced needs to be a step-up high-voltage conversion ratio by using a boost converter based on the necessity of the load [247,248]. Even if the traditional boost converter can, in theory, boost the supply voltage, achieving such a high-voltage conversion ratio would probably direct the boost converter to work at its maximum duty cycle [249,250]. This is not an appropriate solution due to the high-voltage rating devices and vast variations in the voltage with little change in the duty cycle [251,252]. For the problem of inductor current ripple, inductance value can be designed according to appliance necessity and might increase the size and cost of the converter. The interleaved arrangement explains why we reduce the current ripple at the input and output ports and reduce the size of the passive components. The interleaved configuration is an identical DC-DC converter connected in parallel with the existing converter [253–256]. This decreases the component current rating and reduces the distortion at the input and output ports. Many existing DC-DC converter configurations are examined and the configurations are shown in Figures 7.1(a)–(f) and 7.2(a)–(f)[257,258]. Table 7.1 tabulates the voltage conversion ratio of the DC-DC converters. These converters are not an appropriate solution to achieve a high-voltage conversion ratio with reducing ripple, low voltage, and current stress across the switch.

$^{*}V_{out} = V_{o}$

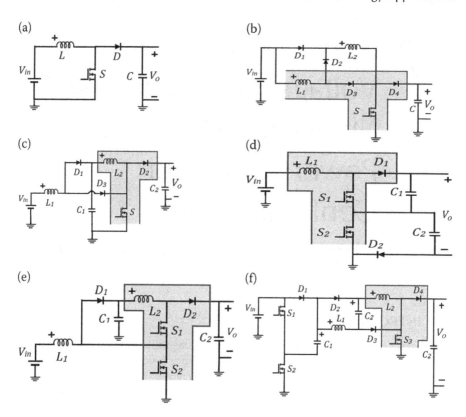

FIGURE 7.1 Power Circuits: (a) Traditional Boost Converter, (b) SI DC-DC Boost Converter [115,116,132], (c) Quadratic Boost Converter with Single Switch [135,106], (d) Traditional Three-Level DC-DC Boost Converter [29,71], (e) Three-Level Quadratic Boost Converter [29,71], (f) Bootstrap Capacitor and Boost Inductor-Based Converter [28,71].

To achieve a high-voltage conversion ratio, a CW-VM-MPC or Nx IMBC configuration is suggested. The complete theoretical background, analysis, design, and working modes of the CW-VM-MPC is discussed in detail. The conspicuous features of CW-VM-MPC configurations are discussed and compared with the existing multistage converter in terms of cost, the number of compliments, and a voltage conversion ratio. We discuss the effect of the internal series resistance of an inductor on the voltage conversion ratio. The simulation results of CW-VM-MPC configurations are provided to verify the functionality of the power converter. The complete set of experimental results of CW-VM-MPC configurations are provided.

7.2 CW-VM-MPC OR NX IMBC CONFIGURATION

The power circuit of the CW-VM-MPC or Nx IMBC configuration is shown in Figure 7.3 [257,258]. The CW-VM-MPC configuration merges the characteristics

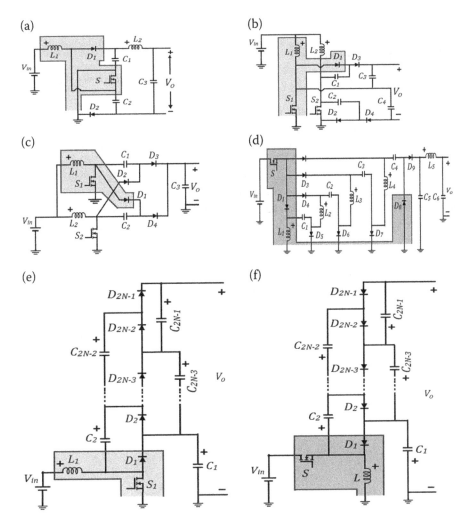

FIGURE 7.2 Power Circuits: (a) Boost Converter Based on SC [71,257,258], (b) Interleaved Two-Phase Quadruple Boost Converter [29,71,257–259], (c) Two-Phase Interleaved High-Voltage Gain Boost Converter Using One Voltage Multiplier Cell [71,257,258], (d) DC-DC Converter with Extra High Voltage [257,258,260], (e) Nx MBC [31,121,203,261], (f) Inverted MBBC [203,204].

of the traditional boost converter interleaved configuration and CW voltage multiplier with *4N-2* diodes, *3N-2* capacitors, 2 identically rated inductors, and 2 control switches are necessary, where N is the output side number of levels.

The CW-VM-MPC offers an N times voltage conversion ratio when compared to the traditional boost converter. The number of stages/levels increases by adding an additional CW voltage multiplier level in the CW-VM-MPC configuration to

TABLE 7.1

Voltage Conversion Ratio of Converters

Power Converter Configuration	Voltage Conversion Ratio Δ = Duty Cycle
Traditional DC-DC Boost Converter [29,71]	$1/(1 - \Delta)$
SI DC-DC Boost Converter [115,116,132]	$(1 + \Delta)/(1 - \Delta)$
Single switch DC-DC Quadratic Boost Converter (QBC) [135,106]	$1/(1 - \Delta)2$
Traditional Three-Level DC-DC Boost Converter [29,71]	$2/(1 - \Delta)$
Three-Level Quadratic Boost Converter [29,71,106]	$1/(1 - \Delta)2$
Bootstrap Capacitor and Boost Inductor-Based Converter [29,71]	$(\Delta/1 - \Delta)+ 3$
Boost Converter Based on the Switched Capacitor [71,257,258]	$(1 + \Delta)/(1 - \Delta)$
Interleaved Two-Phase Quadruple Boost Converters [29,71,257–259]	$4/(1 - \Delta)$
Two-Phase Interleaved High-Voltage Gain Boost Converter Using Voltage Multiplier Cell [71,257,258]	$(VMC + 1)/(1 - \Delta)$
DC-DC Converter with Extra High Voltage [257,258,260]	$4/(1 - \Delta)$
Nx Multilevel Boost Converter (Nx MBC) [31,121,203,261]	$N/(1 - \Delta)$
Multilevel Buck Boost Converter (MBBC) [203,204]	$-(\Delta + (N - 1)/(1 - \Delta))$

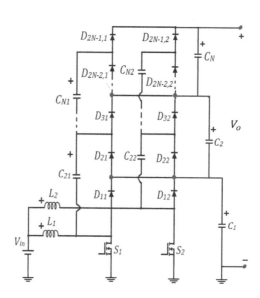

FIGURE 7.3 Power Circuit of CW-VM-MPC Configuration or Nx IMBC.

augment the voltage conversion ratio without disconnecting the primary circuit. The graph of the number of required inductors, diodes, capacitors, and switches versus many levels is depicted in Figure 7.4(a)–(d), respectively. The output voltage and voltage across the capacitor can be seen in equations (7.1) and (7.2).

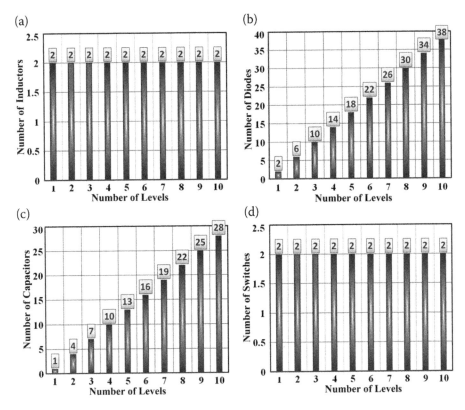

FIGURE 7.4 Component Requirements: (a) Number of Inductors, (b) Number of Diodes, (c) Number of Capacitors, (d) Number of Switches.

$$V_{CN} + V_{CN-1} + \ldots\ldots\ldots + V_{C1} = V_0 \tag{7.1}$$

$$\sum_{r=1}^{N} V_{Cr} = V_0 \tag{7.2}$$

$$\left.\begin{array}{l} \text{Number of Capacitor} = -2 + 3N \\ \text{Number of Inductor} = 2 \\ \text{Number of Diode} = 2(2N - 1) \\ \text{Number of Control Switch} = 2 \end{array}\right\} \tag{7.3}$$

7.2.1 3x CW-VM-MPC CONFIGURATION—WORKING MODES

To give details of the working modes of configuration, 3x CW-VM-MPC is considered with ideal devices or components, the capacitors are high enough, and the

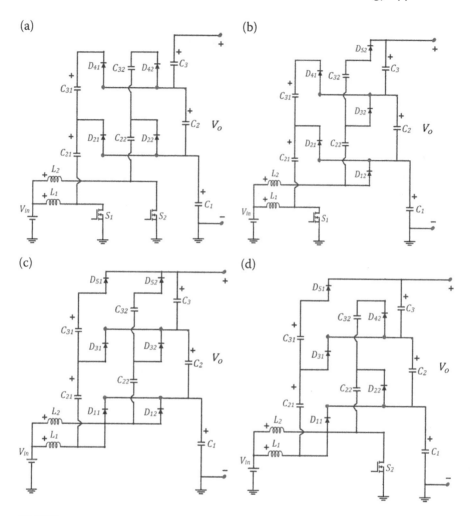

FIGURE 7.5 Equivalent Circuits of the 3x CW-VM-MPC: (a) When Switches S_1 and S_2 Are Turned ON, (b) When Switch S_1 Is Turned ON and Switch S_2 is Turned OFF, (c) When Switches S_2 and S_1 Are Turned OFF, (d) When Switch S_1 Is Turned OFF and Switch S_2 Is Turned ON.

converter is operating in continuous conduction mode (CCM). The CW-VM-MPC configuration can work in four different working modes.

When both control switches (S_2 and S_1) are turned ON, both inductors (L_1 and L_2) are magnetized from the input supply source, and the capacitors (C_1, C_2, and C_3) transferred their energy to the load. At the same time, capacitor voltages make the diodes (D_{22}, D_{21}, D_{42}, and D_{41}) forward-biased and the capacitors (C_1, C_2, and C_3) at the load side discharge via the path of diodes to charge the C_{22}, C_{21}, C_{32}, and C_{31} capacitors. The equivalent circuit of the 3x CW-VM-MPC for this mode is depicted in Figure 7.5(a). In this mode, the D_{12}, D_{11}, D_{32}, D_{31}, D_{52}, and D_{51} diodes are reversed-biased.

The relation between capacitor voltage and output voltage of the 3x CW-VM-MPC when switches (S_2 and S_1) are turned ON can be obtained in equation (7.4).

$$\left. \begin{array}{l} L_2\frac{di_{L2}}{dt} = L_1\frac{di_{L1}}{dt} = V_{in} \\ V_{C_{21}} = V_{C_1} = V_{C_{22}} \\ V_{C_{21}} + V_{C_{31}} = V_{C_1} + V_{C_2} = V_{C_{22}} + V_{C_{32}} \\ V_{C_1} + V_{C_3} + V_{C_2} = V_o \end{array} \right\} \tag{7.4}$$

The equivalent circuit of the 3x CW-VM-MPC when switch S_1 is turned ON and switch S_2 is turned OFF is shown in Figure 7.5(b). The inductor L_1 is magnetized with input source voltage and the C_1, C_3, and C_2 capacitors discharge and charge within the switch S_1 ON time. When C_1, C_3, and C_2 are discharged via the load, this energy is transferred to capacitors C_{21} and C_{31} of the CW voltage multiplier. In this mode, diodes D_{31}, D_{11}, D_{22}, D_{42}, and D_{51} are reversed-biased and diodes D_{41}, D_{21}, D_{32}, D_{52}, and D_{12} are forward-biased. The relation of capacitor voltage and output voltage of 3x CW-VM-MPC when switch S_1 is turned ON and switch S_2 is turned OFF can be seen as equation (7.5).

$$\left. \begin{array}{l} L_1\frac{di_{L1}}{dt} = V_{in},\ V_{C_{21}} = V_{C_1} \\ V_{C_1} + V_{C_2} = V_{C_{21}} + V_{C_{31}},\ V_{C_1} = -L_2\frac{di_{L2}}{dt} + V_{in} \\ V_{C_2} + V_{C_1} = -L_2\frac{di_{L2}}{dt} + V_{C_{22}} + V_{in},\ V_o = V_{C_2} + V_{C_3} + V_{C_1} = -L_2\frac{di_{L2}}{dt} + V_{C_{32}} + V_{C_{22}} \\ \qquad + V_{in} \end{array} \right\}$$

$$(7.5)$$

The equivalent circuit of the 3x CW-VM-MPC when switches S_2 and S_1 are turned OFF is depicted in Figure 7.5(c). In this mode, inductors L_1 and L_2 are demagnitzed. Capacitors C_3, C_1, and C_2 are charged by a series arrangement of L_1, C_{21}, V_{in}, and C_{31} and a series combination of V_{in}, L_2, C_{22}, and C_{32}. Diodes D_{31}, D_{11}, D_{12}, D_{51}, D_{52}, and D_{32} are forward-biased and diodes D_{41}, D_{21}, D_{42}, and D_{22} are reversed-biased. The relation for capacitor voltage and output voltage of 3x CW-VM-MPC when switches S_2 and S_1 are turned OFF can be seen in equation (7.6).

$$\left. \begin{array}{l} V_{C_1} = -L_1\frac{di_{L1}}{dt}V_{in} = -L_2\frac{di_{L2}}{dt}V_{in} \\ V_{C_2} + V_{C_1} = V_{in} + V_{C_{21}} - L_1\frac{di_{L1}}{dt} = V_{in} + V_{C_{22}} - L_2\frac{di_{L2}}{dt} \\ V_{C_2} + V_{C_3} + V_{C_1} = V_{in} + V_{C_{21}} - L_1\frac{di_{L1}}{dt} + V_{C_{31}} \\ V_{C_2} + V_{C_3} + V_{C_1} = V_{in} + V_{C_{22}} - L_2\frac{di_{L2}}{dt} + V_{C_{32}} = V_o \end{array} \right\} \tag{7.6}$$

The equivalent circuit of the 3x CW-VM-MPC when switch S_1 is turned OFF and switch S_2 is turned ON is shown in Figure 7.5(d). The inductor L_2 is magnetized with an input source voltage and C_1, C_2, and C_3 capacitors discharge within the switch S_2 ON time and switch S_1 OFF time. When C_2, C_3, and C_1 discharge via load, tis also charges the C_{32} and C_{22} capacitors of the CW voltage multiplier. In this mode, diodes D_{31}, D_{11}, D_{22}, D_{42}, and D_{51} are forward-biased and diodes D_{41}, D_{21}, D_{32}, D_{52}, and D_{12} are reversed-biased. The relation of capacitor voltage and output voltage of 3x CW-VM-MPC when switch S_1 is turned OFF and switch S_2 is ON can be seen in equation (7.7).

$$\left.\begin{aligned}
&L_2\frac{di_{L2}}{dt} = V_{in}, \ V_{C_{22}} = V_{C_1}, \ V_{C_{32}} + V_{C_{22}} = V_{C_2} + V_{C_1} \\
&V_{C_1} = -L_1\frac{di_{L1}}{dt} + V_{in}, \ V_{C_1} + V_{C_2} = V_{in} + V_{C_{21}} - L_2\frac{di_{L2}}{dt} \\
&V_{C_2} + V_{C_3} + V_{C_1} = V_{in} + V_{C_{31}} + V_{C_{21}} - L_2\frac{di_{L2}}{dt} = V_o
\end{aligned}\right\} \qquad (7.7)$$

7.2.2 THE EFFECT OF THE INDUCTOR EQUIVALENT SERIES RESISTANCE ON THE CW-VM-MPC

For real-time purposes, the voltage conversion ratio of any step-up converter is constrained by resistance of devices and passive components in a particular inductor of the converter. Here, R_{L2} and R_{L1} are in equivalent series resistance with inductors L_2 and L_1, respectively. For simplicity, the switch and diode drop are neglected in the analysis of the CW-VM-MPC configuration with load R_o. The mathematical expression for the CW-VM-MPC configuration can be seen in equations (7.8)–(7.13):

$$\frac{V_{in}}{I_{in}^{-1}} = \frac{V_o}{I_o^{-1}} = V_o^2 \times \frac{1}{R_o} = NV_o \times \frac{V_{C1}}{R_o} = \frac{N^2 V_{in} V_{C1}}{(1 - \Delta)} \times \frac{1}{R_o} \qquad (7.8)$$

$$I_{in} = I_{L2} + I_{L1} = \frac{V_{C1} \times N^2}{\bar{\Delta}} \times \frac{1}{R_o} \qquad (7.9)$$

When switches S_1 and S_2 are turned ON,

$$\left.\begin{aligned}
&-I_{L1} \times R_{L1} + V_{in} = L_1 \times \frac{di_{L1}}{dt} \\
&-I_{L2} \times R_{L2} + V_{in} = L_2 \times \frac{di_{L2}}{dt}
\end{aligned}\right\} \qquad (7.10)$$

When switch S_1 turned ON and S_2 is turned OFF,

$$\left.\begin{aligned}
&-I_{L1} \times R_{L1} + V_{in} = L_1 \times \frac{di_{L1}}{dt} \\
&-I_{L2} \times R_{L2} - V_{C_1} + V_{in} = L_2 \times \frac{di_{L2}}{dt}
\end{aligned}\right\} \qquad (7.11)$$

When switches S_1 and S_2 are turned OFF,

$$\left.\begin{array}{l} -I_{L1} \times R_{L1} - V_{C_1} + V_{in} = L_1 \times \frac{di_{L1}}{dt} \\ -I_{L2} \times R_{L2} - V_{C_1} + V_{in} = L_2 \times \frac{di_{L2}}{dt} \end{array}\right\} \quad (7.12)$$

When switch S_1 is turned OFF and S_2 is turned ON,

$$\left.\begin{array}{l} -I_{L1} \times R_{L1} - V_{C_1} + V_{in} = L_1 \times \frac{di_{L1}}{dt} \\ -I_{L2} \times R_{L2} + V_{in} = L_2 \times \frac{di_{L2}}{dt} \end{array}\right\} \quad (7.13)$$

In a steady-state condition, the average inductor voltage is 0. Consider Δ as the duty ratio of gate pulses given to switches S_1 and S_2 with equal switching frequency. The pulse provides to switch S_2 a delay of 50% compared to a pulse of switch S_1. According to the inductor-volt-second-balance method, the equation for inductor voltages can be seen in (7.14).

$$\left.\begin{array}{l} \Delta(V_{in} - I_{L1} \times R_{L1}) + \bar{\Delta}(V_{in} - I_{L1} \times R_{L1} - V_{C1}) = V_{L1} = 0 \\ \Delta(V_{in} - I_{L2} \times R_{L2}) + \bar{\Delta}(V_{in} - I_{L2} \times R_{L2} - V_{C1}) = V_{L2} = 0 \end{array}\right\} \quad (7.14)$$

By solving (7.14),

$$\left.\begin{array}{l} I_{L1} \times R_{L1} + \bar{\Delta} \times V_{C1} = V_{in} \\ I_{L2} \times R_{L2} + \bar{\Delta} \times V_{C1} = V_{in} \end{array}\right\} \quad (7.15)$$

$$V_{in} = \frac{I_{L2}R_{L2} + I_{L1}R_{L1}}{2} + \bar{\Delta} \times V_{C1} \quad (7.16)$$

When we use identical and similarly rated inductors, we assume the internal resistance of both inductors is the same ($R_{L2} = R_{L1} = R_L$). We can write the mathematical expression for converters as equations (7.17)–(7.20). The voltage conversion ratio can be seen in equation (7.21).

$$V_{in} = \frac{R_L (I_{L1} + I_{L2})}{2} + \bar{\Delta} \times V_{C1} \quad (7.17)$$

$$I_{in} = I_{L2} + I_{L1} \quad (7.18)$$

$$V_{in} = \frac{R_L I_{in}}{2} + \bar{\Delta} \times V_{C1} \quad (7.19)$$

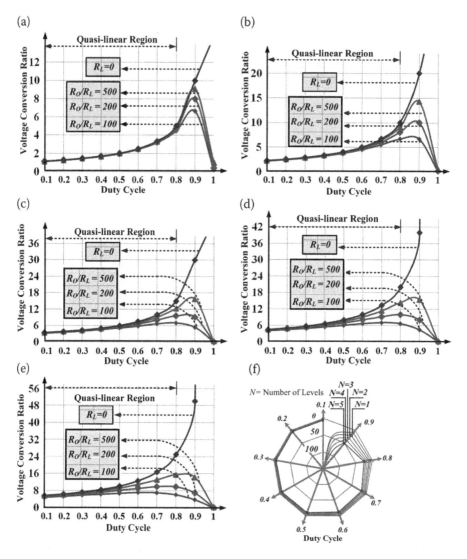

FIGURE 7.6 Plot of Voltage Conversion Ratio Versus Duty Cycle: (a) When $N = 1$, (b) When $N = 2$, (c) When $N = 3$, (d) When $N = 4$, (e) When $N = 5$, (f) Voltage Conversion Ratio with $R_{esrL} = 0$.

$$\left. \begin{array}{l} V_{in} = \dfrac{1}{2} \times \dfrac{N^2 \times V_{C1} \times R_L}{\bar{\Delta} \times R_o} + \bar{\Delta} \times V_{C1} \\[2mm] V_{in} = \dfrac{1}{2} \times \dfrac{N \times V_o \times R_L}{\bar{\Delta} \times R_o} + \dfrac{\bar{\Delta} \times V_o}{N} \end{array} \right\} \qquad (7.20)$$

$$\frac{V_o}{V_{in}} = \frac{1}{\dfrac{1}{2} \dfrac{N \times R_L}{\bar{\Delta} \times R_o} + \dfrac{1}{\Delta N}} = \frac{N}{\bar{\Delta} + \dfrac{1}{2} \dfrac{N^2 \times R_L}{\bar{\Delta} \times R_o}} \qquad (7.21)$$

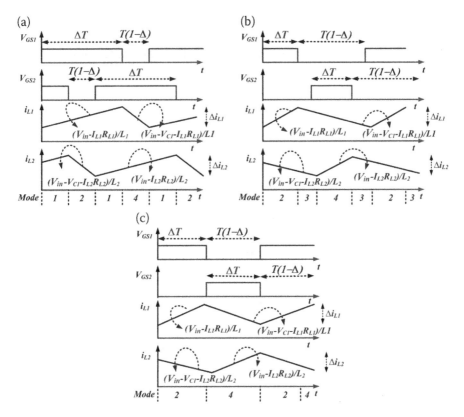

FIGURE 7.7 Inductor Current Waveforms: (a) When D or $\Delta > 50\%$, (b) When D or $\Delta < 50\%$, (c) When D or $\Delta = 50\%$.

The graph of the voltage conversion ratio against the duty cycle (Δ) is plotted for various cases of R_L/R_o when $N = 1$ to 5 and shown in Figure 7.6(a)–(e). A separate graph is plotted for $N = 1$ to 5 in one graph with $R_{esrL} = 0$. The voltage conversion ratio is rising linearly in the quasi-linear region, up to 80% of the duty cycle.

The slope of the current of the inductors (L_1 and L_2) is examined for three ranges of duty cycles: 1) a duty cycle larger than 50%, 2) a duty cycle smaller than 50%, and 3) a duty cycle equal to 50%. These are shown in Figure 7.7(a)–(c), respectively. We find that the CW-VM-MPC configuration operates in modes 2, 4, and 1 when the duty cycle is more than 50%, the CW-VM-MPC configuration operates in mode 4, 3, and 2 when the duty cycle is smaller than 50%, and the CW-VM-MPC configuration operates only in modes 4 and 2 when the duty cycle is equal to 50%. In mode 1, the L_2 and L_1 inductors' current slopes are positive. Thus, the magnetization of inductors L_2 and L_1 takes place in mode 1. In mode 2, the L_2 and L_1 inductors' current slopes are negative and positive, respectively. Therefore, in mode 2, the magnetization of inductor L_1 and demagnetization of inductor L_2 has occurred. In mode 3, the L_2 and L_1 inductors' current slopes are negative. As a

TABLE 7.2

Inductor Charging and Discharging State with the Slope of Current

Duty Cycle of Both Switches	Modes of Suggested Converters				Positive Slope	Negative Slope
	Mode 1	Mode 2	Mode 3	Mode 4		
Greater than 50%	L_1: Charging	L_1: Charging	Does not occur	L_1: Discharging	$(V_{in} - I_{L1}R_{L1})/L_1$	$(V_{in} - VC_1 - I_{L1}R_{L1})/L_1$
	L_2: Charging	L_2: Discharging	Does not occur	L_2: Charging	$(V_{in} - I_{L2}R_{L2})/L_2$	$V_{in} - V_{C1} - I_{L2}R_{L2})/L_2$
Less than 50%	Does not occur	L_1: Charging	L_1: Discharging	L_1: Discharging	$(V_{in} - I_{L1}R_{L1})/L_1$	$(V_{in} - VC_1 - I_{L1}R_{L1})/L_1$
	Does not occur	L_2: Discharging	L_2: Discharging	L_2: Charging	$(V_{in} - I_{L2}R_{L2})/L_2$	$V_{in} - VC1 - I_{L2}R_{L2})/L_2$
Equal to 50%	Does not occur	L_1: Charging	Does not occur	L_1: Discharging	$(V_{in} - I_{L1}R_{L1})/L_1$	$(V_{in} - VC_1 - I_{L1}R_{L1})/L_1$
	Does not occur	L_2: Discharging	Does not occur	L_2: Charging	$(V_{in} - I_{L2}R_{L2})/L_2$	$V_{in} - VC_1 - I_{L2}R_{L2})/L_2$

result, demagnetization of inductors L_2 and L_1 takes place in mode 3. In mode 4, the L_2 and L_1 inductors' current slopes are positive and negative, respectively. In mode 4, demagnetization of L_1 inductor and magnetization of L_2 inductor are taking place; the charging and discharging condition of inductors L_1 and L_2 are given in Table 7.2.

7.3 THE CW-VM-MPC CONFIGURATION WITH RECENT DC-DC CONVERTERS

In Table 7.3, the CW-VM-MPC configuration is compared with recent DC-DC converters in terms of the voltage conversion ratio. The comparison is also shown graphically in Figure 7.8. We find that the 4x CW-VM-MPC provides a superior voltage conversion ratio when compared to recent DC-DC converters at a given duty cycle. As a result, the CW-VM-MPC offers a workable solution to step up the voltage with immense value for photovoltaic applications. In Table 7.4, the voltage across the switch in the CW-VM-MPC is compared with existing DC-DC converters; it is found that the switch voltage is very low in the CW-VM-MPC configuration. Therefore, low-rating components are appropriate to design a CW-VM-MPC configuration.

TABLE 7.3
Voltage Conversion Ratio of Converters

Converter	Duty Cycle				
	0.5	0.6	0.7	0.8	0.9
Traditional DC-DC Boost Converter [29,71]	2.00	2.50	3.33	5.00	10.00
SI DC-DC Boost Converter [115,116,132]	3.00	4.00	5.67	9.00	19.00
Single Switch DC-DC Quadratic Boost Converter (QBC) [135,106]	4.00	6.25	11.11	25.00	100.00
Traditional Three-Level DC-DC Boost Converter [29,71]	4.00	5.00	6.67	10.00	20.00
Three-Level Quadratic Boost Converter [29,71,106]	4.00	6.25	11.11	25.00	100.00
Bootstrap Capacitor and Boost Inductor-Based Converter [29,71]	7.00	9.00	12.33	19.00	39.00
Boost Converter Based on the SC [71,257,258]	3.00	4.00	5.67	9.00	19.00
Interleaved Two-Phase Quadruple Boost Converter [29,71,257–259]	8.00	10.00	13.33	20.00	40.00
Two-Phase Interleaved High-Voltage Gain Boost Converter Using Voltage Multiplier Cell [71,257,258]	4.00	5.00	6.67	10.00	20.00
DC-DC Converter with Extra High Voltage [257,258,260]	8.00	10.00	13.33	20.00	40.00
4x MBC [31,121,203,261]	8.00	10.00	13.33	20.00	40.00
4x MBBC [204]	–7.00	–9.00	–12.33	–19.00	–39.00
4x IMBC or CW-VM-MPC	8.00	10.00	13.33	20.00	40.00

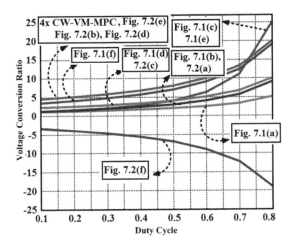

FIGURE 7.8 CW-VM-MPC Configuration Is Compared with Recently Addressed DC-DC Converters in Terms of Voltage Conversion Ratio.

7.4 VALIDATION OF THE 3X CW-VM-MPC

The CW-VM-MPC configuration is simulated for a three-level (3x CW-VM-MPC or 3x IMBC) configuration in Matrix Laboratory 9.0 (R2016a) with the designed

TABLE 7.4

Voltage Stress Comparison of 4x IMBC with Recent Converters

Converter Type	Voltage Across Switch
Traditional DC-DC Boost Converter [29,71]	V_o
SI DC-DC Boost Converter [115,116,132]	V_o
Single Switch DC-DC Quadratic Boost Converter (QBC) [106,135]	V_o
Traditional Three-Level DC-DC Boost Converter [29,71]	$V_o/2$
Three-Level Quadratic Boost Converter [29], [71,106]	$V_o(1 - \Delta),$ $V_o - Vo(1 - \Delta)$
Bootstrap Capacitor and Boost Inductor-Based Converter [29,71]	Vo
Boost Converter Based on the SC [71,257,258]	$V_o/(1 - \Delta)$
Interleaved Two-Phase Quadruple Boost Converter [29,71,257–259]	$V_o/4$
Two-Phase Interleaved High-Voltage Gain Boost Converter Using Voltage Multiplier Cell [71,257,258]	$V_o/2$
DC-DC Converter with Extra High Voltage [257,258,260]	$V_o/4$
4x MBC [31,121,203, 261]	V_o/N or $V_{in}/(1 - \Delta)$
4x MBBC [204]	$Vin/(1 - \Delta)$
4x IMBC or CW-VM-MPC	V_o/N or $V_{in}/(1 - \Delta)$

TABLE 7.5

Simulation Parameters

Parameters	Value
Output Voltage and Input Voltage	120 V/10 V
Load Resistance, Power Load	144 Ω, 100 W
Number of Designed Levels of the Converter	3
Capacitance, Inductance	220 µF, 150 µH
Switching Frequency, Duty Cycle (Δ)	50 kHz, 0.75

parameter to verify the concept and functionality of the converter configuration. Table 7.5 depicts the simulation parameters used to design the CW-VM-MPC configuration. Figure 7.9(a) depicts the waveforms of output and input voltage of the 3x CW-VM-MPC configuration. We find that the voltage conversion ratio of 3x CW-VM-MPC is 12 at a duty cycle of 75%. Thus, the required output voltage of 120 V is obtained from the 10-V input voltage. Figure 7.9(b) depicts the voltage at various output levels. We find that every level of converter contributes to equal voltage ($V_o/3$, i.e., 40 V). Thus, 40 V is measured the first level, 80 V is noted at the second level, and 120 V is noted at the third level.

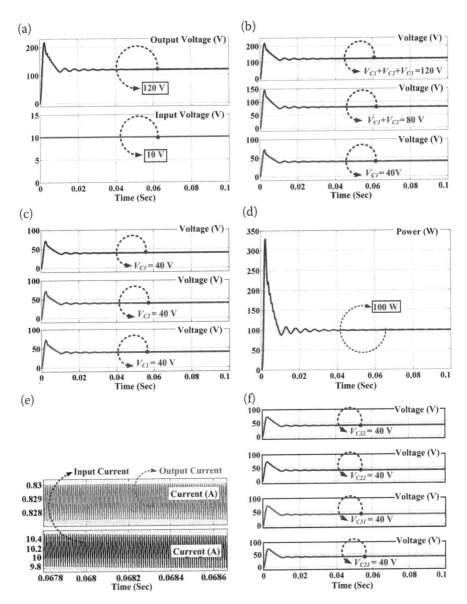

FIGURE 7.9 Simulation Results: (a) Output and Input Voltage (b) Voltage at Various Output Level (c) Shows the Voltage Sharing of Output Side Capacitors (C_2, C_1 and C_3) (d) Power Graph of 3x CW-VM-MPC Converter (e) Ripple of Output and Input Current (f) Voltage Sharing Across Capacitors of Voltage Multiplier.

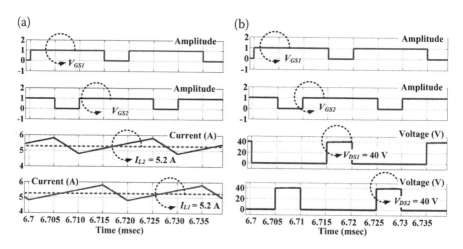

FIGURE 7.10 Simulation Results: (a) Inductor Current Waveform, (b) Voltage Across the Switch.

TABLE 7.6
Detail of Hardware Components or Semiconductor Devices

Parameters	Value
Switches (S_1 and S_2)	IRF540
Power Diodes	BYQ28E
Power Inductors	8 A, 150 uH
Capacitors	50 V, 220 uF
Motor Driver IC	IC 293D

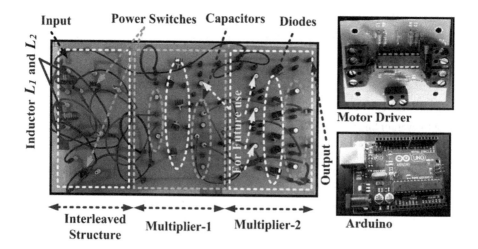

FIGURE 7.11 Hardware Prototype Model of Nx CW-VM–MPC or Nx IMBC.

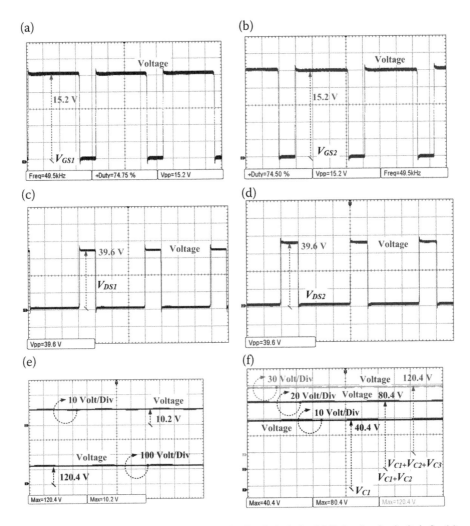

FIGURE 7.12 Experiment Results: (a) Pulse for Switch S_1, (b) Pulse for the Switch S_2, (c) Voltage Across Switch S_1, (d) Voltage Across Switch S_2, (e) Input and Output Voltage, (f) Voltage Waveform at Various Output Levels.

Figure 7.9(c) shows the voltage sharing of output-side capacitors (C_2, C_1, and C_3). It is observed that the voltage sharing of output-side capacitors is equal to $V_o/3$ (i.e., 40 V). The voltage sharing across all of the output-side capacitors (C_1, C_2, and C_3) shows that the converter performs satisfactorily without unbalancing. Figure 7.9(d) depicts the power graph of the 3x CW-VM-MPC converter. The 3x CW-VM-MPC converter provides 100 W of power and the ripple of output and input current is depicted in Figure 7.9(e). It is observed that the ripple content in input current and output current is 600 mA and 2 mA, respectively. Thus, the 3x CW-VM-MPC configuration provides a low input and output current ripple. Figure 7.9(f) shows the

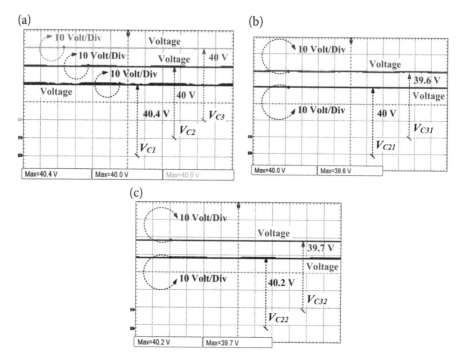

FIGURE 7.13 Experiment Results: (a) Voltage Across Capacitor, (b) Voltage Across Capacitors C_{31} and C_{21}; (c) Voltage Across Capacitors C_{32} and C_{22}.

voltage sharing across capacitors of the voltage multiplier. It is observed that voltage across multiplier capacitors is equal to $V_o/3$ (i.e., 40 V). The voltage sharing across all the multiplier capacitors shows that the converter performs satisfactorily.

Figure 7.10(a) depicts the current waveform of the inductor with a switch gate pulse. We observe that 5.2 A of current is flowing through the inductors (L_2 and L_1), which are exactly half of the input current. Figure 7.10(b) depicts the drain-to-source voltage of switches (V_{DS2} and V_{DS1}) with gate-to-source voltage of switches (V_{GS2} and V_{GS1}). The drain-to-source voltage of both switches is equal to $V_o/3$ (i.e., 40 V). This drain-to-source voltage of the switch is always the same, even with the addition of the voltage levels at the output.

The CW-VM-MPC (Nx IMBC) configuration is experimentally tested for three levels, and the result always shows a good agreement with the simulation results and theoretical approach. The detail of hardware components used for the hardware prototype is shown in Table 7.6.

The hardware prototype of the 3x CW-VM-MPC (3x IMBC) configuration is depicted in Figure 7.11. The Arduino Uno is utilized to generate pulses for power switches and a motor driver IC293D is used as driver IC. This motor IC can operate two switches at the same time and gives 15.2 V at the output terminal when directly fed with the pulse generated by the Arduino Uno that has a magnitude of 5 V.

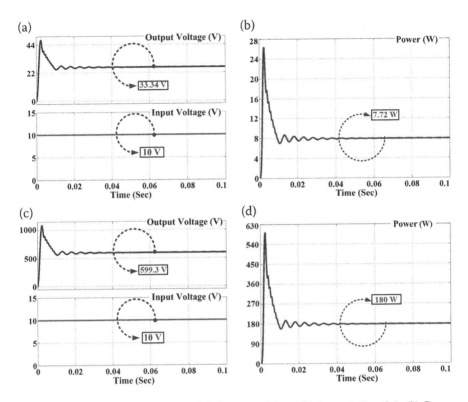

FIGURE 7.14 Simulation Results: (a) Output and Input Voltage at $D = 0.1$, (b) Power Graph of 3x CW-VM-MPC Converter at $D = 0.1$, (c) Output and Input Voltage at D = 0.95, (d) Power Graph of 3x CW-VM-MPC Converter at D = 0.95.

Figure 7.12(a)–(b) shows the pulses for switches S_1 and S_2 (V_{GS1} and V_{GS2}), respectively. Switches S_1 and S_2 operate with a 75% duty cycle and 50-kHz switching frequency. Figure 7.12(c)–(d) depicts the voltage across power switches S_1 and S_2, respectively. We find that voltage across switches S_2 and S_1 is 39.6 V and 39.6 V, respectively. Figure 7.12(e) depicts the waveform of output and input voltages. We find that 120.4 V is achieved at the output port of the 3x CW-VM-MPC configuration by using an input supply of 10 V. Figure 7.12(f) depicts the voltage at various levels of the 3x CW-VM-MPC configuration.

Experimentally, 40.4 V is achieved at the first level, 80.4 V is achieved at the second level, and 120.4 V is achieved at the third level. Figure 7.13(a) shows the voltage sharing of output-side capacitors. We observe that the voltage across capacitors C_1, C_2, and C_3 is 40.4 V, 40 V, and 40 V, respectively. Figure 7.13(b)–(c) depicts the voltage of multiple capacitors: C_{21} and C_{31} and C_{22} and C_{32}, respectively. We observe that the voltage across capacitors C_{21}, C_{31}, C_{22}, and C_{32} is 40 V, 39.6 V, 40.2 V, and 39.7 V, respectively.

We observe that 33.34 V is achieved at the output when the input is 10 V and the output power is 7.72 W. At the duty cycle, $D = 0.95$ and load = 2000 Ω; the obtained output voltage and obtained power is shown in Figure 7.14(c)–(d), respectively. Notice that 599.3 V is achieved at the output when input is 10 V, and output power is 180 W. Therefore, the voltage conversion ratio of the converter is 3.33 and 60 at $D = 0.1$ and $D = 0.95$, respectively. The results matched the theoretical analysis.

7.5 CONCLUSION

A Cockcroft Walton (CW) MPC configuration or Nx IMBC is articulated for photovoltaic applications. The CW-VM-MPC configuration combines the feature of CW, a traditional boost converter, and an interleaved structure. The CW-VM-MPC converter provides a high-voltage conversion ratio (N time compared to traditional boost converter) with reduced voltage/current ripple and less switch stress. The striking features of the CW-VM-MPC configuration are positive output, high-voltage conversion ratio without the use of a transformer, less current and voltage ripples, less voltage across the switch, and ease in increasing the number of levels or stages. Higher reliability is compared to converters that require several numbers of switches. The CW-VM-MPC concept is validated through hardware implementation and simulation. Both simulation and experimental results always show agreement with the theoretical approach.

8 Conclusion and Future Direction

In this book, we proposed and presented a series of new, non-isolated, unidirectional DC-DC multistage power converter configurations for renewable energy applications with hardware implementation and investigation studies. At the beginning of the book, a review of power electronics-based photovoltaic systems was presented. Based on the review of the photovoltaic systems, we found that the photovoltaic systems were influenced by the chosen power electronics converter configurations and the MPPT control. Also, we found that photovoltaic energy will play an essential role in any available or future power system made by power electronics technology. Because of the low DC output of photovoltaic cells or panels, a DC-DC converter with a high-voltage conversion ratio capability is essential for real-time applications or before feeding energy to a grid via an inverter. Therefore, the DC-DC power converter is the most vital constituent in the photovoltaic power conversion system. MPPT is also necessary to extract maximum power from the photovoltaic source. Thus, to select an excellent MPPT control, a review with the merits and demerits of each algorithm was presented. Next, to choose the most suitable technique to attain a high-voltage conversion ratio, a complete review of existing unidirectional, non-isolated DC-DC multistage power converters was presented, and we found that not all of them can convert the low voltage into high voltage. Thus, most of the configurations are not suitable for photovoltaic energy applications. Therefore, four DC-DC multistage power converter configurations were suggested with high-voltage conversion capabilities by employing an original arrangement of reactive elements and semiconductor devices.

A breed of buck-boost converters called the X-Y power converter family was articulated and designed by using a concept of SI and voltage lift switches. The features of the X-Y power converter family include a single switch configuration, a smaller number of stages, a high-voltage conversion ratio without a transformer, and less voltage stress. Based on the number of stages, the whole X-Y power converter family is divided into three configurations: two-stage X-Y power converter configuration, three-stage X-Y power converter configuration, N-stage L-Y power converter configuration. The complete theoretical analysis was provided, and simulation results verified the concept. Hardware implementation and simulation results of a self-balanced DC-DC multistage power converter configuration without magnetic components for photovoltaic applications was presented. PIC controller and TLP250 driver were used to generating the pulses for the control switches. Suggested self-balanced converter configurations were compared, with the existing converter approach, and we found that the self-balanced converter configuration offers a practical solution to advance the photovoltaic systems in terms of modularity, control, and cost. The noticeable features of the self-balanced configuration are lack of magnetic components, input current is continuous, ratings of semiconductor devices are low, modular approachs and levels can increase to increase the voltage conversion ratio, and only two switches needed. The self-balanced configuration finds DC-link photovoltaic applications in DC-AC systems where voltage balancing of the capacitor is the major challenge.

To boost the voltage with a high conversion ratio, a T-SC MPC configuration was suggested with hardware implementation. A T-SC multistage converter combines the features of the transformer, switch capacitor, and the conventional boost converter. Compared with the existing transformer and coupled inductor converter, we found that the T-SC multistage converter offers a high-voltage conversion ratio with a smaller number of components and devices. The features of the T-SC MPC include a high-voltage conversion ratio, the input current is continuous, only one switch is needed, and the source voltage across the switch is low (only one inductor and one transformer are required). The complete hardware model of a CW-VM-MPC is implemented to reduce the voltage and current ripples and to obtain a high-voltage conversion ratio. Arduino Uno is used to generating pulses for power switches and motor driver IC293D driver IC. The CW-VM-MPC converter combines the features of CW, a traditional boost converter, and an interleaved structure. The CW-VM-MPC configuration provides a positive output with a high-voltage conversion ratio without the use of a transformer. The CW-VM-MPC converter was compared with the existing converter, and we found that the CW-VM-MPC converter configuration provides a viable solution for a photovoltaic system to achieve a high-voltage conversion ratio. Both simulation and experimental results always agree with the theoretical approach.

For the future task, this book opens the research direction with several motivating tasks such as:

- A complete closed-loop hardware realization of recommended converter configurations for an electric drive system using an advance controller like the TMS320F2812 DSP controller or dSPACE.
- Improvement of the MPPT control to extract maximum power from the photovoltaic cell.
- A complete closed-loop Implementation of the T-SC MPC configuration with voltage multiplier by using the MPPT and advance controllers like TMS320F2812 DSP or dSPACE.
- Design more DC-DC configurations by hybridization of available boosting techniques like SIs and SCs.
- Reduce voltage multiplier stages in a CW-VM-MPC configuration by using an SC and SI concept.

The suggested configurations in Chapters 4 to 7 can be found in applications like electrical vehicles (EVs), fuel cell vehicles (FCVs), wind energy systems, and more.

References

1. S. S. Williamson, A. K. Rathore, F. Musavi, "Industrial electronics for electric transportation: Current state-of-the-art and future challenges", *IEEE Trans. Ind. Electron.*, vol. 62, no. 5, pp. 3021–3032, May 2015.
2. J. P. Ribau, C. M. Silva, J. M. C. Sousa, "Efficiency, cost and life cycle CO_2 optimization of fuel cell hybrid and plug-in hybrid urban buses", *Appl. Energy*, vol. 129, pp. 320–335, Sep. 2014.
3. T. Ida, K. Murakami, M. Tanaka, "A stated preference analysis of smart meters, photovoltaic generation, and electric vehicles in Japan: Implications for penetration and GHG reduction", *Energy Res. Soc. Sci.*, vol. 2, pp. 75–89, June 2014.
4. U. S. Energy Information Administration (EIA), "International Energy Outlook", *U.S. Department of Energy*, Washington, DC: Office of Energy Analysis, Report no.: DOE/EIA-0484(2013), 2013.
5. F. Querini, S. Dagostino, S. Morel, P. Rousseaux, "Greenhouse gas emissions of electric vehicles associated with wind and photovoltaic electricity", *Energy Procedia*, vol. 20, pp. 391–401, June 2012.
6. International Energy Agency, "Global EV Outlook—Beyond One Million Electric Cars", 2016.
7. International Energy Agency, "Energy Technology Perspectives 2012, IEA, Paris", https://www.iea.org/reports/energy-technology-perspectives-2012, 2012.
8. J. Y. Yong, V. K. Ramachandaramurthy, K. M. Tan, N. Mithulananthan, "A review on the state-of-the-art technologies of electric vehicle, its impacts and prospects", *Renew. Sustain. Energy Rev.*, vol. 49, pp. 365–385, Sep. 2015.
9. B. Türkay, A. Y. Telli, "Economic analysis of standalone and grid connected hybrid energy systems", *Renew. Energy*, vol. 36, pp. 1931–1943, July 2011.
10. S. Padmanaban, M. Bhaskar, P. Maroti, F. Blaabjerg, V. Fedák, "An original transformer and switched-capacitor (T & SC)-based extension for DC-DC boost converter for high-voltage/low-current renewable energy applications: Hardware implementation of a new T & SC boost converter", *Energies*, vol. 11, no. 4, p. 783, April 2018.
11. S. G. Chalk, J. F. Miller, "Key challenges and recent progress in batteries, fuel cells, and hydrogen storage for clean energy systems", *J. Power Sources*, vol. 159, no. 1, pp. 73–80, Sep. 2006.
12. Energy Information Administration (EIA), "International energy outlook 2009", *United States Dept of Energy*, May 2009. [Online].
13. International Energy Agency (IEA), "World energy outlook 2012", 2012.
14. EPIA, "Global market outlook for photo-voltaic until 2013", *European Photovoltaic Industry Association*, 2009. [Online].
15. G. R. Timilsina, L. Kurdgelashvili, P. A. Narbel, "Solar energy: Markets, economics and policies", *Renew. Sustain. Energy*, vol. 16, no. 1, pp. 449–465, Jan. 2012.
16. F. Blaabjerg, Y. Yang, Ke Ma, X. Wang, "Power Electronics—The Key Technology for Renewable Energy System Integration", Conf. Proc. of IEEE 4th International conference on renewable Energy Research and Application, IEEE-ICRERA'15, Palermo, (Italy), pp. 1618–1626, November 22–25, 2015.
17. F. Blaabjerg, Ke Ma, Y. Yang, "Power Electronics for Renewable Energy Systems –Status and Trends", Conf. Proc. of IEEE Intl. Conf. on Integrated Power System, IEEE-CIPS'14, Nuremberg, (Germany), pp. 1–11, February 25–27, 2014.

18. A. S. Joshi, I. Dincer, B. V. Reddy, "Performance analysis of photovoltaic systems: A review". *Renew. Sustain. Energy Rev.*, vol. 13, no. 8, pp. 1884–1897, Oct. 2009.

19. R. Teodorescu, M. Liserre, P. Rodriguez, *"Grid Converters for Photovoltaic and Wind Power Systems"*, Hoboken, NJ, USA: Wiley, 2011.

20. U.S. Energy Information Administration, "International Energy Outlook Transportation Sector Energy Consumption", Washington DC, Sep. 2017.

21. P. Sanjeevikumar, F. Blaabjerg, P. Wheeler, J. Olorunfemi Ojo, A. Ertas, "High-voltage DC-DC converter topology for PV energy utilization-investigation and implementation", *J. Electr. Power Comp. Syst.,* vol. 45, pp. 221–232, Dec. 2016.

22. R. R. Ahrabi, H. Ardi, M. Elmi and A. Ajami, "A Novel Step-Up Multiinput DC-DC Converter for Hybrid Electric Vehicles Application", *IEEE Trans. Power Electron.*, vol. 32, no. 5, pp. 3549–3561, May 2017.

23. W. Li, X. He, "Review of non isolated high-step-up dc/dc converters in photovoltaic grid-connected applications", *IEEE Trans. Ind. Electron.*, vol. 58, no. 4, pp. 1239–1250, April 2011.

24. W. Kleinkauf, G. Cramer, M. Ibrahim, *"PV Systems Technology-State of the Art Developments and Trends in Remote Electrification"*, Fountain Valley, CA, (USA): SMA Technologies AG, 2005.

25. B. Sri-Revathi, M. Prabhakar, "Non isolated high gain DC-DC converter topologies for PV applications—A comprehensive review", *Renew. Sustain. Energy Rev.*, vol. 66, pp. 920–933, Dec. 2016.

26. M. Forouzesh, Y. Siwakoti, S. Gorji, F. Blaabjerg, B. Lehman, "Step-up DC–DC converters: A comprehensive review of voltage-boosting techniques, topologies, and applications", *IEEE Trans. Power Electron.*, vol. 32, no. 12, pp. 9143–9178, Dec. 2017.

27. J. C. Rosas Caro, M. Fernando, J. Mayo-Maldonado, J. M. Gonzalez, L. Halida, E. Jejus, "Transformer-less high gain boost converter with input current ripple cancelation at a selectable duty cycle", *IEEE Trans. Ind. Electron.*, vol. 60, no. 10, pp. 4492–4499, Oct. 2013.

28. W. Li, X. Lv, Y. Deng, J. Liu, X. He, "A Review of Non-Isolated High Step-Up DC/DC Converters in Renewable Energy Applications", Conf. Proc, IEEE 24th Annual Applied Power Electronics Conf. and Exposition, IEEE-APEC'09, Washington DC (USA), pp. 364–369, February 15–19, 2009.

29. F. L. Tofoli, W. Josias de Paula, D. de, S. Oliveira Júnior, D. de, C. Pereira, "Survey on non-isolated high-voltage step-up dc–dc topologies based on the boost converter", *IET Power Electron.*, vol. 8, no. 10, pp. 2044–2057, Oct. 2015.

30. M. S. Bhaskar, M. Meraj, A. Iqbal, S. Padmanaban, "Nonisolated symmetrical inter-leaved multilevel boost converter with reduction in voltage rating of capacitors for high-voltage microgrid applications", *IEEE Trans. Ind. Appl.*, vol. 55, no. 6, pp. 7410–7424, Nov. 2019.

31. A. Iqbal, M. S. Bhaskar, M. Meraj, S. Padmanaban, "DC-transformer modelling, analysis and comparison of the experimental investigation of a non-inverting and non-isolated Nx multilevel boost converter (Nx MBC) for low to high DC voltage applications", *IEEE Access*, vol. 6, pp. 70935–70951, 2018.

32. P. Sanjeevikumar, G. Grandi, P. Wheeler, F. Blaabjerg, J. Loncarski, "A Simple MPPT Algorithm for Novel PV Power Generation system by High Output Voltage DC-DC Boost Converter", Conf. Proc. 24th IEEE Intl. Symposium on Ind. Electronics, Rio de Janeiro, Brazil, pp. 214–220, June 3–5, 2015.

33. M. Forouzesh, Y. Siwakoti, S. Gorji, F. Blaabjerg, B. Lehman, "A Survey on Voltage Boosting Techniques for Step-up DC-DC Converters", Conf. Proc. IEEE Energy Conversion Congress and Exposition, IEEE-ECCE'16, Milwaukee (USA), pp. 1–6, September 18–22, 2016.

34. M. S. Bhaskar, D. J. Almakhles, S. Padmanaban, J. B. Holm-Nielsen, A. R. Kumar, S. O. Masebinu, "Triple-mode active-passive parallel intermediate links converter with high voltage gain and flexibility in selection of duty cycles", *IEEE Access*, vol. 8, pp. 134716–134727, 2020.

35. B. Chandrasekar, C. Nallaperumal, P. Sanjeevikumar, M. S. Bhaskar, J. Holm-Nielsen, Z. Leonowicz, S. Masebinu, "Non-isolated high-gain triple port DC–DC multilevel converter with positive output voltage for photovoltaic applications", *IEEE Access*, vol. 8, pp. 113649–113666, 2020.

36. K. I. Hwu, C. F. Chuang, W. C. Tu, "High voltage boosting converter based on boost strap capacitors and boost inductors", *IEEE Trans. Ind. Electron.*, vol. 60, no. 6, pp. 2178–2193, June 2013.

37. V. Das, P. Sanjeevikumar, K. Venkitusamy, R. Selvamuthukumaran, F. Blaabjerg, P. Siano, "Recent advances and challenges of fuel cell based power system architectures and control— A review", *Renew. Sustain. Energy Rev.*, vol. 73, pp. 10–18, June 2017.

38. F. Blaabjerg, F. Iov, T. Kerekes, R. Teodorescu, "Trends in Power Electronics and Control of Renewable Energy Systems", Conf. Proc. 14th IEEE International Power Electronics and Motion Control Conference, IEEE-EPE-PEMC'10, Metropol Lake Resort Ohrid, (Macedonia), pp. k-1–k-19, September 6–8, 2010.

39. F. Li, C. Li, J. Shi, J. Zhao, X. Yang, Z. Chen, "Evaluation Index system for photovoltaic system statistical characteristics under hazy weather conditions in central China", *IET Renew. Power Gen.*, vol. 11, no. 14, pp. 1794–1803, Dec. 2017.

40. A. Raghoebarsing, K. Anand, "Performance and economics analysis of a 27 kW grid-connected photovoltaic system in Suriname", *IET Renew. Power Gen.*, vol. 11, no. 12, pp. 1545–1554, Nov. 2017.

41. P. G. Nikhil, D. Subhakar, "Approaches for developing a regression model for sizing a stand-alone photovoltaic system", *IEEE J. Photovolt.*, vol. 5, no. 1, pp. 250–257, Jan. 2015.

42. P. G. Bueno, J. G. Hernandez, F. J. Ruiz-Rodriguez, "Stability assessment for transmission system with large utility-scale photovoltaic units", *IET Renew. Power Gen.*, vol. 10, no. 5, pp. 584–597, April 2016.

43. Y. Hu, W. Xiao, W. Cao, B. Ji, D. John Morrow, "Three-Port DC-DC converter for stand-alone photovoltaic system", *IEEE Trans. Power Electron.*, vol. 30, no. 6, pp. 3068–3076, June 2015.

44. M. Islam, S. Mekkhilef, "H6-type transformerless single-phase inverter for grid-tied photovoltaic system", *IET Power Electron.*, vol. 8, no. 4, pp. 636–644, April 2015.

45. M. Calais, V. Agelidis, "Multilevel converters for single-phase grid connected photovoltaic systems, an overview", in Proc. of IEEE - Intl. Symposium on Industrial Electronics, ISIE'08, pp. 224–229, 1998.

46. N. Jenkins, "Photovoltaic systems for small-scale remote power supplies", *Power Eng. J.*, vol. 9, no. 2, pp. 89–96, April 1995.

47. M. Abella and F. Chenlo, "Choosing the right inverter for gridconnected PV systems", *Renew. Energy World*, vol. 7, no. 2, pp. 132–147, March–April 2004.

48. S. B. Kjaer, J. K. Pedersen, F. Blaabjerg, "A review of singlephase grid connected inverters for photovoltaic modules", *IEEE Trans. Ind. Appl.*, vol. 41, no. 5, pp. 1292–1306, Sep. 2005.

49. M. S. Bhaskar, S. Padmanaban, F. Blaabjerg, "A multistage DC-DC step-up self-balanced and magnetic component-free converter for photovoltaic applications: Hardware implementation", *Energies*, vol. 10, no. 5, Art. no. 5, May 2017.

50. L. Tang, W. Xu, C. Mu, " Analsyis for step-size optimisation on MPPT algorithm for photovoltaic systems", *IET Power Electron.*, vol. 10, no. 13, pp. 1647–1654, Nov. 2017.

51. J. Ahmed, Z. Salam, "An accurate method for MPPT to detect the partial shading occurrence in a PV system", *IEEE Trans. Ind. Inform.*, vol. 13, no. 5, pp. 2151–2161, May 2017.

52. S. B. Jeaprabha, A. I. Selvakumar, " Model based MPPT for shaded and mismatched modules of phototvoltaic farm", *IEEE Trans. Sustain. Energy*, vol. 8, no. 4, pp. 1763–1771, Oct. 2017.

53. K. Kumar, N. Ramesh babu, K. R. Prabhu, "Design and analysis of RBFN-based single MPPT controller for hybrid solar and wind energy system", *IEEE Access*, vol. 5, pp. 15308–15317, Aug. 2017.

54. K. Chen, S. Tian, Y. Cheng, L. Bai, "An improved MPPT controller for photovoltaic system under partial sading conditiion", *IEEE Trans. Sustain. Energy*, vol. 5, no. 3, pp. 978–985, July 2014.

55. R. Jiang, Y. Han, S. Zhang, "Wide-range, high-precision and low-comlexixty MPPT circuit based on pertrub and observe algorithm", *Electron. Lett.*, vol. 53, no. 16, pp. 1141–1142, Aug. 2017.

56. K. L. Lian, J. H. Jhang, I. S. Tian, "A maximum power point tracking method based on perturb and observe combined with particle swarm optimization", *IEEE J. Photovolt.*, vol. 4, no. 2, pp. 626–633, March 2014.

57. B. Subudhi, R. Pradhan, "A comparative study on maximum power point tracking techniques for photovoltaic power systems", *IEEE Trans. Sustain. Energy*, vol. 4, no. 1, pp. 89–98, Jan. 2013.

58. S. Mohanty, B., Subudhi, P. V. Ray, "A grey wolf-assisted perturb and observe MPPT algorithm for a PV system", *IEEE Trans. Energy Convers.*, vol. 32, no. 1, pp. 340–347, March 2017.

59. D. Sera, L. Mathe, T. Kerekes, S. V. Spataru, R. Teodorescu, "On the perturb and observe and icremental conductance MPPT method for PV system", *IEEE J. Photovolt.*, vol. 3, no. 3, pp. 1070–1078, July 2013.

60. N. Zakzouk, M. A. Elsaharty, A. K. Abdelsalam, A. A. Helal, B. W. Williams, "Improved performance low-cost incremental conductance PV MPPT technique", *IET Renew. Power Gen.*, vol. 10, no. 4, pp. 561–574, March 2016.

61. G. Hsieh, H. Hsieh, C. Tsai, C. Wang, "Photovoltaoic power-increment-aided incremental-conductance MPPT with two phased tracking", *IEEE Trans. Power Electron.*, vol. 28, no. 6, pp. 2895–2911, June 2013.

62. A. W. Leedy, L. Guo, K. A. Aganah, "A Constant Voltage MPPT Method for a Solar Powered Boost Converter with DC Motor Load", Conf. Proc. of IEEE Southeastcon, Orlando, (USA), pp. 1–6, March 15–18, 2012.

63. K. A. Aganah, A. W. Leedy, "A constant voltage maximum power point tracking method for solar powered system", IEEE 43rd Southeastern Symposium on System Theory, Auburn, USA, pp. 125–130, March 14–16, 2011.

64. M. A. Masoum, H. Dehbhoei, E. F. Fuchs, " Theortical and experimental alaysis of photovoltaic systems with voltage and current-based maximum power point tracking", *IEEE Power Eng. Rev.*, vol. 22, no. 8, pp. 62–62, 2002.

65. P. K. Maroti, R. Al-Ammari, M. S. Bhaskar, M. Meraj, A. Iqbal, P. Sanjeevikumar, S. Rahman, "New tri-switching state non-isolated high gain DC–DC boost converter for microgrid application", *IET Power Electron.*, vol. 12, no. 11, pp. 2741–2750, 2019.

66. G. Zhang, Z. Wang, H. Iu, C. Si-zhe, Y. Ye, B. Zhang, Y. Zhang, "Unique modular structure of multicell high-boost converters with reduced component currents", *IEEE Trans. Power Electron.*, vol. 33, no. 9, pp. 7795–7804, Sep. 2018.

67. N. Gupta, M. S. Bhaskar, D. Almakhles, P. Sanjeevikumar, F. Blaabjerg, Z. Leonowicz, "Two-Tier Converter: A New Structure of High Gain DC-DC Converter with Reduced Voltage Stress", IEEE Intl. Conf. on Environment and Electrical Engineering and IEEE

Ind. and Commercial Power Systems Europe, (EEEIC / I CPS Europe), pp. 1–6, Madrid, Spain, June 9–12, 2020.

68. G. Mademlis, G. K. Steinke, Alfred Rufer, "Feed forward based control in a a DC-DC converter of asymmetric multistage-stacked boost architecture", *IEEE Power Electron.*, vol. 32, no. 2, pp. 1507–1517, Feb. 2017.

69. J. Chen, D. Sha, Y. Yan, Bin Lu, X. Liao, "Cascaded high voltage conversion ratio bidirectinal nonsiolated DC-DC converter with variable switching frequecny", *IEEE Trans. Power Electron.*, vol. 33, no. 2, pp. 1399–1409, Feb. 2018.

70. S. K. Pidaparthy, B. Choi, H. Kim, Y. Kim, "Stabilizing effetcs of load subsystem on multistage DC-to-DC power conversion system", *IEEE J. Emerging Select. Top. Power Electron.*, vol. 5, no. 4, pp. 1589–1603, Dec. 2017.

71. M. S. Bhaskar, V. Ramachandaramurthy, P. Sanjeevikumar, F. Blaabjerg, D. Ionel, M. Mitolo, D. Almakhles, "Survey of DC-DC non-isolated topologies for unidirectional power flow in fuel cell cehicles", *IEEE Access*, pp. 1–37, 2020.

72. M. S. Bhaskar, M. Meraj, A. Iqbal, S. Padmanaban, P. K. Maroti, R. Alammari, "High gain transformer-less double-duty-triple-mode DC/DC converter for DC microgrid", *IEEE Access*, vol. 7, pp. 36353–36370, 2019.

73. J. Ali, D. Almakhles, S. A. Ibrahim, S. Alyami, S. Selvam, M. S. Bhaskar, "A generalized multilevel inverter topology with reduction of total standing voltage", *IEEE Access*, vol. 8, pp. 168941–168950, 2020.

74. M. S. Bhaskar, D. Almakhles, P. Sanjeevikumar, D. Ionel, F. Blaabjerg, J. He, R. Kumar, "Investigation of a transistor clamped T-type multilevel H-bridge inverter with inverted double reference single carrier PWM technique for renewable energy applications", *IEEE Access*, vol. 8, pp. 161787–161804, 2020.

75. N. Gupta, M. S. Bhaskar, D. Almakhles, P. Sanjeevikumar, S. Umashankar, Z. Leonowicz, M. Mitolo, "Novel Non-Isolated Quad-Switched Inductor Double-Switch Converter for DC Microgrid Application", IEEE Intl. Conf. on Environment and Electrical Engineering and IEEE Ind. and Commercial Power Systems Europe, (EEEIC / I CPS Europe), pp. 1–6, Madrid, Spain, June 9–12, 2020.

76. S. De, G. Swathika, N. Tewari, A. Venkatesan, S. Umashankar, M. S. Bhaskar, P. Sanjeevikumar, Z. Leonowicz, M. Mitolo, "Implementation of designed PV integrated controlled converter system", *IEEE Access*, vol. 8, pp. 100905–100915, 2020.

77. N. Priyadarshi, P. Sanjeevikumar, J. B. Holm-Nielsen, F. Blaabjerg, M. S. Bhaskar, "An experimental estimation of hybrid ANFIS–PSO-based MPPT for PV grid integration under fluctuating sun irradiance", *IEEE Syst. J.*, vol. 14, no. 1, pp. 1218–1229, March 2020.

78. M. Lakshmi, S. Hemamalini, "Nonisolated high gain DC–DC converter for DC microgrids", *IEEE Trans. Ind. Electron.*, vol. 65, no. 2, pp. 1205–1212, Feb. 2018.

79. M. S. Bhaskar, R. Alammari, M. Meraj, S. Padmanaban, A. Iqbal, "A new triple-switch-triple-mode high step-up converter with wide range of duty cycle for DC microgrid applications", *IEEE Trans. Industry Appl.*, vol. 55, no. 6, pp. 7425–7441, Nov. 2019.

80. H. Zhang, S. Wang, Y. Li, Q. Wang, D. Fu, "Two capacitor transformer winding capacitance models for common mode EMI noise analaysis in isolated DC-DC converters", *IEEE Trans. Power Electron.*, vol. 32, no. 11, pp. 8458–8469, Nov. 2017.

81. T. Liang, J. Lee, S. Chen, J. Chen, L. Yang, "Novel isolated high-step-up DC-DC converter with voltage lift", *IEEE Trans. Ind. Electron.*, vol. 60, no. 4, pp. 1483–19491, April 2013.

82. A. Emrani, E. Adib, H. Farzanehfard, " Single switch soft switched isolated DC-DC converter", *IEEE Trans. Power Electron.*, vol. 27, no. 4, pp. 1952–1957, April 2012.

83. K. Park, G. Woo Moon, M. Joong Youn, "Non isolated high step-up stacked converter based on boost integrated isolated converter", *IEEE Trans. Power Electron.*, vol. 26, no. 2, pp. 577–587, Feb. 2011.

84. S. Lee, H. Do, "Isolated SEPIC DC-DC converter with ripple-free input current and lossless snubber", *IEEE Trans. Ind. Electron.*, vol. 65, no. 2, pp. 1254–1262, Feb. 2018.

85. Y. Shi, H. Li, "Isolated modular multilevel DC-DC converter with DC fault furrent control capacibility based on current-fed dual active bridge for MVDC application", *IEEE Trans. Power Electron.*, vol. 33, no. 3, pp. 2145–2161, March 2018.

86. D. G. Bandeira, I. Barbi, "A T-type isolated zero-voltage switching DC_DC converter with capacitve output", *IEEE Trans. Power Electron.*, vol. 32, no. 6, pp. 4210–4218, June 2017.

87. S. A. Gorji, M. Ektesabi, J. Zheng, "Isolated switched-boost push pull DC_DC converter for step-up application", *Electron. Lett.*, vol. 53, no. 3, pp. 177–179, Feb. 2017.

88. M. Nguyen, Y. Lim, J. Cjoi, G. Cho, "Isolated high step-up DC-DC converter based on quasi-switched-boost network", *IEEE Trans. Ind. Electron.*, vol. 63, no. 12, pp. 7553–7562, Dec. 2016.

89. R. Erickson, D. Maksimovib, *"Fundamentals of Power Electronics"*, 2nd Edition, United States: Springer, 2001.

90. M. H. Rashid, *"Power Electronics Handbook-Devices, Circuits and Applications"*, 3rd Edition, Burlington, USA: Elsevier, 2011.

91. B. Bryant, M. K. Kazimierczuk, "Derivation of the cuk pwm DC-DC converter circuit topology", Conf. Proc, IEEE Intl. Symposium on Circuits and Systems, IEEE-ISCAS'03, vol. 3, pp. 292–295, Bangkok (Thailand), May 25–28, 2003.

92. B. Bryant, M. K. Kazimierczuk, "Derivation of the buck-boost pwm DC-DC converter circuit topology", Conf. Proc, IEEE Intl. Symposium on Circuits and Systems, IEEE-ISCAS'02, Phoenix-Scottsdale (USA), vol. 5, pp. 841–844, May 26–29, 2002.

93. M. S. Bhaskar, P. K. Maroti, D. K. Prabhakar, "Novel topological derivations for DC-DC converters", *IJCEM Int. J. Comput. Eng. Manag.*, vol. 16, no. 6, pp. 49–53, Nov. 2013.

94. T. G. Wilson, "The evolution of power electronics", *IEEE Trans. Power Electron.*, vol. 15, no. 3, pp. 439–446, May 2000.

95. B. Axelrod, Y. Berkovich, A. Ioinovici, "Hybrid switched-capacitor Cuk/Zeta/Sepic converters in step-up mode", Conf. Proc, IEEE Intl. Symposium Circuits and System, IEEE-ISCAS'05, Kobe (Japan), pp. 1310–1313, May 23–26, 2005.

96. A. Bratcu, I. Munteanu, S. Bacha, D. Picault, B. Raison, "Cascaded DC-DC converter photovoltaic systems: Power optimization issues", *IEEE Trans. Ind. Electron.*, vol. 58, no. 2, pp. 403–411, Feb. 2011.

97. G. R. Walker, P. C. Sernia, "Cascaded DC-DC converter connection of photovoltaic modules", *IEEE Trans. Power Electron.*, vol. 19, no. 4, pp. 1130–1139, July 2004.

98. M. S. Bhaskar, S. Padmanaban, F. Blaabjerg, P. W. Wheeler, "An Improved Multistage Switched Inductor Boost Converter (Improved M-SIBC) for Renewable Energy Applications: A key to Enhance Conversion Ratio", IEEE 19th Workshop on Control and Modeling for Power Electronics (COMPEL), pp. 1–6, Padua, Italy, June 25–28, 2018.

99. M. Ortiz-Lopez, J. Leyva-Ramos, E. Carbajal-Gutierrez, J. Morales-Saldana, "Modeling and analysis of switch-mode cascade converters with a single active switch", *IET Power Electron.*, vol. 1, no. 4, pp. 478–487, Dec. 2008.

100. R. D. Middlebrook, "Transformer-less DC-to-DC converters with large conversion ratios", *IEEE Trans. Power Electron.*, vol. 3, no. 4, pp. 484–488, Oct. 1988.

101. P. Maroti, P. Sanjeevikumar, M. S. Bhaskar, F. Blaabjerg, P. Wheeler, "New Inverting Modified CUK Converter Configurations with Switched Inductor (MCCSI) for High-Voltage/Low-Current Renewable Applications", 20th European Conf. on Power Electron. and Appl. (EPE'18 ECCE Europe), pp. 1–10. Riga, Latvia.

102. P. K. Maroti, M. S. Bhaskar, S. Padmanaban, J. B. Holm-Nielsen, T. Sutikno, A. Iqbal, "A New Multilevel Member of Modified CUK Converter Family for Renewable

Energy Applications", IEEE Conf. on Energy Conversion (CENCON), pp. 224–229, Oct. 2019, Yogyakarta, Indonesia, October 16–17, 2019.

103. P. Yang, J. Xu, G. Zhou, S. Zhang, "A new quadratic boost converter with high voltage step-up ratio and reduced voltage stress", Conf. Proc., IEEE 7th Intl. Power Electronics and Motion Control Conf., IEEE-PEMC'12, pp. 1164–1168, June 2–5, 2012.

104. D. S. Wijeratne, "Quadratic power conversion for power electronics: Principles and circuits", *IEEE Trans. Circuits Syst.-I*, vol. 59, no 2, pp. 426–438, Feb. 2012.

105. Y. Yuan-mao, K. Cheng, "Quadratic boost converter with low buffer capacitor stress", *IET Power Electron.*, vol. 7, no. 5, pp. 1162–1170, May 2014.

106. Y. Novaes, A. Rufer, I. Barbi, "A new quadratic, three-level, DC/DC converter suitable for fuel cell applications", Conf. Proc, IEEE Power Conversion Conf., IEEE-PCC'07, Nagoya (Japan), pp. 601–607, April 2–5, 2007.

107. S. Li, Y. Zheng, B. Wu, K. Smedley, "A family of resonant two-switch boosting switched-capacitor converters with ZVS operation and a wide line regulation range", *IEEE Trans. Power Electron.*, vol. 33, no. 1, pp. 448–459, January 2018.

108. B. Axelrod, Y. Berkovich, A. Ioinovici, "Switched–capacitor (SC) switched-inductor (SL) structures for getting hybrid step-down Cuk/Zeta/Sepic converters", IEEE Intl. Symposium on Circuits and Systems, IEEE-ISCAS'06, Kos Island (Greece), pp. 5063–5066, May 21–24, 2006.

109. G. Palumbo, D. Pappalardo, "Charge pump circuits: An overview on design strategies and topologies", *IEEE Circuits Syst. Mag.*, vol. 10, no. 1, pp. 31–45, March 2010.

110. M. Seeman, S. Sanders, "Analysis and optimization of switched capacitor DC-DC converters", *IEEE Trans. Power Electron.*, vol. 23, no. 2, pp. 841–851, March 2008.

111. M. Seeman, "A design methodology for switched-capacitor DC-DC converters", *Electrical Engineering and Computer Science*, University of California, Berkeley, USA, Tech. Rep. no. UCB/EECS-2009-78, May 2009.

112. F. Luo, "Investigation of Switched-capacitorized DC/DC converters", Conf. Proc, IEEE 6th Intl. Power Electronics Motion Control Conf., IEEE-PEMC'09, Wuhan (China), pp. 1074–1079, May 17–20, 2009.

113. D. Zhou, A. Pietkiewicz, S. Cuk, "A three-switch high-voltage converter", *IEEE Trans. Power Electron.*, vol. 14, no. 1, pp. 177–183, Jan. 1999.

114. D. Navamani, K. Vijayakumar, R. Jegatheeesanm, A. Lavanya, "High step-up DC-DC converter by switched inductor and voltage multiplier cell for automotive applications", *J. Electr. Eng. Technol.*, vol. 11, pp. 1921–1935, April 2016.

115. B. Axelrod, Y. Berkovich, A. Ioinovici, "Switched capacitor/switched-inductor structures for getting transformer less hybrid DC-DC PWM converters", *IEEE Trans. Circuits Syst. I*, vol. 55, no. 2, pp. 687–696, March 2008.

116. A. Ioinovici, "Switched-capacitor power electronics circuits", *IEEE Circuits Syst. Mag.*, vol. 1, no. 1, pp. 37–42, Jan. 2001.

117. Y. Berkovich and B. Axelrod, "Switched coupled-inductor cell for DC–DC converters with very large conversion ratio", *IET Power Electron.*, vol. 4, no. 3, pp. 309–315, March 2011.

118. Y. Wang, H. Yin, S. Han, A. Alsabbagh, C. Ma, " A Novel Switched Capacitor Circuit for Battery Cell Balancing Speed Improvement", IEEE 26th Intl. Symposium on Industrial Electronics, IEEE-ISIE'17, Edinburgh, Scotland, pp. 1977–1982, June 19–21, 2017.

119. E. Ismail, M. Al-Saffar, A. Sabzali, A. Fardoun, "A family of single-switch PWM converters with high step-up conversion ratio", *IEEE Trans, Circuits Syst.-I*, vol. 55, no. 4, pp. 1159–1171, May 2008.

120. J. Rosas-Caro, J. Ramirez, F. Peng, A. Valderrabano, "A DC-DC multilevel boost converter", *IET Power Electron.*, vol. 3, no. 1, pp. 129–137, Jan. 2010.

121. A. Farooq, Z. Malik, Z. Sun, G. Chen, "A review of non-isolated high step-down DC-DC converters", *Int. J. Smart Home*, vol. 9, no. 8, pp. 133–150, Jan. 2015.

122. W. Liou, M. Yeh, Y. Kuo, "A high efficiency dual-mode buck converter, IC for portable applications", *IEEE Trans. Power Electron.*, vol. 23, no. 2, pp. 667–677, March 2008.

123. O. Lee, S. Cho, G. Moon, "Interleaved buck converter having low switching losses and improved step-down conversion ratio", *IEEE Trans. Power Electron.*, vol. 27, no. 8, pp. 3664–3675, August 2012.

124. E. Carbajal, J. Morales-Saldan, A. Ramos, "Modeling of a single-switch quadratic buck converter", *IEEE Trans. Aerosp. Electron. Syst.*, vol. 41, no. 4, pp. 1450–1456, Oct. 2005.

125. A. Ayachit, M. Kazimierczuk, "Steady-State Analysis of PWM Quadratic Buck Converter in CCM", Conf. Proc, IEEE 56th IEEE Intl. Midwest Symposium on Circuits and Systems, IEEE-MWSCAS'13, Columbus (USA), pp. 49–52, August 4–7, 2013.

126. J. Morales-Saldana, J. Leyva-Ramos, E. Carbajal-Gutiérrez, M. Ortiz-Lopez, "Average current-mode control scheme for a quadratic buck converter with a single switch", *IEEE Trans. Power Electron.*, vol. 23, no. 1, pp. 485–490, Jan. 2008.

127. F. Sa, C. Eiterer, D. Ruiz-Caballero, S. Mussa, "Double Quadratic Buck Converter", Conf. Proc, IEEE Power Electronics Conf., IEEE-COBEP'13, Gramado (Brazil), pp. 36–43, October 27–31, 2013.

128. S. Galateanu, "Triple Step-Down DC-DC Converters", Conf. Proc, IEEE 27th Annual Power Electronics Specialists Conference, IEEE-PESC'96, Baveno, (Italy), pp. 408–413, June 23–27, 1996.

129. S. Xiong, S. C. Tan, S. C. Wong, "Analysis and design of a high voltage-gain hybrid switched-capacitor buck converter", *IEEE Trans. Circuits Syst. - I: Regular Papers*, vol. 59, no. 5, pp. 1132–1141, May 2012.

130. Y. Jiao, F. Luo, "N-Switched-Capacitor buck converter: Topologies and analysis", *IET Power Electron.*, vol. 4, no. 3, pp. 332–341, March 2011.

131. N. Muntean, O. Cornea, O. Pelan, C. Lascu, "Comparative Evaluation of Buck and Hybrid Buck DC-DC Converters for Automotive Applications", Conf. Proc, IEEE 15th Intl. Power Electronics and Motion Control Conference, EPE-PEM'12, Novi Sad (Serbia), pp. 1–6, September 4–6, 2012.

132. S. Sadaf, M. S. Bhaskar, M. Meraj, A. Iqbal, N. Alemadi, "A novel modified switched inductor boost converter with reduced switch voltage stress", *IEEE Trans. Ind. Electron.*, pp. 1–1, 2020.

133. P. K. Maroti, P. Sanjeevikumar, M. S. Bhaskar, M. Meraj, A. Iqbal, R. Al-Ammari, "High gain three-state switching hybrid boost converter for DC microgrid applications", *IET Power Electron.*, vol. 12, no. 14, pp. 3656–3667, 2019.

134. L. Yang, T. Liang, J. Chen, "Transformer-less DC-DC converters with high step-up voltage gain", *IEEE Trans. Ind. Electron.*, vol. 56, no. 8, pp. 3144–3152, Aug. 2009.

135. Y. Ye, K. Cheng, "Survey stress", *IET Power Electron.*, vol. 7, no. 5, pp. 1162–1170, May 2014.

136. M. Prudente, L. Pfitscher, G. Emmendoerfer, E. Romaneli, R. Gules, "Voltage multiplier cells applied to non-isolated DC-DC converters", *IEEE Trans. Power Electron.*, vol. 23, no. 2, pp. 871–887, March 2008.

137. A. Tomaszuk, A. Krupa, "High efficiency high step-up DC/DC converters—a review", *Bull. Polish Acad. Sci.: Tech. Sci.*, vol. 59, no. 4, pp. 475–483, Jan. 2011.

138. A. Fardoun, E. Ismail, "Ultra step-up DC-DC converter with reduced switch stress", *IEEE Trans. Ind. Appl.*, vol. 46, no. 5, pp. 2025–2034, Sep. 2010.

139. Y. Jiao, F. Luo, M. Zhu, "Voltage-lift-type switched-inductor cells for enhancing DC–DC boost ability: Principles and integrations in Luo converter", *IET Power Electron.*, vol. 4, no. 1, pp. 131–142, Jan. 2011.

140. O. Cornea, O. Pelan, N. Muntean, "Comparative Study of Buck and Hybrid Buck Switched Inductor DC-DC Converters", Conf. Proc, IEEE 13th Intl. Conf. on Optimization of Electrical and Electronic Equipment, IEEE-OPTIM'12, Brasov (Romania), pp. 853–858, May 24–26, 2012.

141. O. Lee, S. Cho, G. Moon, "Interleaved buck converter having low switching losses and improved step-down conversion ratio", *IEEE Trans. Power Electron.*, vol. 27, no. 8, pp. 3664–3675, Aug. 2012.

142. E. Carbajal, J. Morales-Saldan, J. L. Ramos, "Modeling of a single-switch quadratic buck converter", *IEEE Trans. Aerosp. Electron. Syst.*, vol. 41, no. 4, pp. 1450–1456, Oct. 2005.

143. A. Ayachit, M. Kazimierczuk, "Steady-State Analysis of PWM Quadratic Buck Converter in CCM", Conf. Proc, IEEE, 56th Intl. Midwest Symposium on Circuits and Systems, IEEE-MWSCAS'13, Columbus (USA), pp. 49–52, August 4–7, 2013.

144. S. Xiong, S. Wong, S. Tan, "A Series of Exponential Step-Down Switched-Capacitor Converters and Their Applications in Two-Stage Converters", Conf. Proc, IEEE Intl. Symposium on Circuits and Systems, IEEE-ISCAS'13, Beijing (China), pp. 701–704, May 19–23, 2013.

145. S. B. Mahajan, R. M. Kulkarni, K. Anita, C. Pooja, "Non isolated switched inductor SEPIC converter topologies for photovoltaic boost applications", Conf. Proc, IEEE Intl. Conf. on Circuit, Power and Computing Technologies, IEEE-ICCPCT'16, pp. 1–6, Nagarcoil (India), 18–19 March 2016.

146. Y. Zhang, C. Zhang, J. Liu, Y. Cheng, "Comparison of conventional dc-dc converter and a family of diode-assisted DC-DC converter", Conf. Proc, IEEE 7th Intl. Power Electronics and Motion Control Conf., IEEE-PEMC'12, Harbin (China), pp. 1718–1723, June 2–5, 2012.

147. Y. Ye, K. Cheng, "A family of single-stage switched-capacitor-inductor PWM converters", *IEEE Trans. Power Electron.*, vol. 28, no. 11, pp. 5196–5205, Nov. 2013.

148. Y. Tang, T. Wang, D. Fu, "Multicell switched-inductor/switched capacitor combined active-network converters", *IEEE Trans. Power Electron.*, vol. 30, no. 4, pp. 2063–2072, April 2015.

149. K. Cheng, Y. Ye, "Duality approach to the study of switched inductor power converters and its higher-order variations", *IET Power Electron.*, vol. 8, no. 4, pp. 489–496, April 2015.

150. Y. Tang, D. Fu, T. Wang, Z. Xu, "Hybrid switched-inductor converters for high step-up conversion", *IEEE Trans. Ind. Electron.*, vol. 62, no. 3, pp. 1480–1490, March 2015.

151. H. Liu, F. Li, "A novel high step-up converter with a quasi active switched-inductor structure for renewable energy systems", *IEEE Trans. Power Electron.*, vol. 31, no. 7, pp. 5030–5039, July 2015.

152. Z. Shi, S. Ho, K. Cheng, "Static performance and parasitic analysis of tapped-inductor converters", *IET Power Electron.*, vol. 7, no. 2, pp. 366–375, Feb. 2014.

153. W. Williams, "Unified synthesis of tapped-inductor DC-to-DC converters", *IEEE Trans. Power Electron.*, vol. 29, no. 10, pp. 5370–5383, Oct. 2014.

154. A. Grant, Y. Darroman, J. Suter, "Synthesis of tapped-inductor switched-mode converters", *IEEE Trans. Power Electron.*, vol. 22, no. 5, pp. 1964–1969, Sep. 2007.

155. Q. Zhao, F. Lee, "High-efficiency, high step-up DC-DC converters", *IEEE Trans. Power Electron.*, vol. 18, no. 1, pp. 65–73, Jan. 2003.

156. R. Wai, R. Duan, "High-efficiency DC/DC converter with high voltage gain", *IEE Proc. Electric Power Appl., IET*, vol. 152, no. 4, pp. 793–802, July 2005.

157. W. Yu, C. Hutchens, J. Lai, J. Zhang, G. Lisi, A. Djabbari, G. Smith, T. Hegarty, "High efficiency converter with charge pump and coupled inductor for wide input photovoltaic AC module applications", IEEE Energy Conversion Congress and Exposition, IEEE-ECCE'09, San Jose, (USA), pp. 3895–3900, September 20–24, 2009.

158. T. Liang, S. Chen, L. Yang, J. Chen, A. Ioinovici, "A single switch boost-flyback DC-DC converter integrated with switched-capacitor cell", Conf. Proc, IEEE 8th Intl. Conf. on Power Electronics and ECCE Asia, IEEE-ICPE-ECCE'11, Jeju (South Korea), pp. 2782–2787, May 30–June 3, 2011.

159. P. Ki-Bum, M. Gun-Woo, Y. Myung-Joong, "High step-up boost converter integrated with a transformer-assisted auxiliary circuit employing quasi-resonant operation", *IEEE Trans. Power Electron.*, vol. 27, no. 4, pp. 1974–1984, April 2012.

160. N. Zhang, D. Sutanto, "High-voltage-gain quadratic boost converter with voltage multiplier", *IET Power Electron.*, vol. 8, no. 12, pp. 2511–2519, Dec. 2015.

161. X. Guo, C. Huang, Y. Xu, W. Lin, "The Nonlinear Control of Tapped Inductor Buck Converter Based on Port-controlled Hamiltonian Model", Conf. Proc, IEEE 33rd Intl. Telecommunications Energy Conference, IEEE-INTELEC'11, Amsterdam (Netherlands), pp. 1–8, 12 Dec. 2011.

162. H. Yi-Ping, C. Jiann-Fuh, L. Tsorng-Juu, Y. Lung-Sheng, "Novel high step-up DC-DC converter for distributed generation system", *IEEE Trans. Ind. Electron.*, vol. 60, no. 4, pp. 1473–1482, April 2013.

163. S. Chen, T. Liang, L. Yang, J. Chen, "A cascaded high step-up DC-DC converter with single switch for microsource applications", *IEEE Trans. Power Electron.*, vol. 26, no. 4, pp. 1146–1153, April 2011.

164. Y. Siwakoti, F. Blaabjerg, P. Loh, G. Town, "High-voltage boost quasi-Z-source isolated DC/DC converter", *IET Power Electron.*, vol. 7, no. 9, pp. 2387–2395, Sep. 2014.

165. Y. Siwakoti, F. Blaabjerg, P. C. Loh, "Quasi-Y-source boost DC-DC converter", *IEEE Trans. Power Electron.*, vol. 30, no. 12, pp. 6514–6519, Dec. 2015.

166. Y. Siwakoti, P. Loh, F. Blaabjerg, G. Town, "Y-source impedance network", *IEEE Trans. Power Electron.*, vol. 29, no. 7, pp. 3250–3254, July 2014.

167. Y. Siwakoti, F. Blaabjerg, P. Loh, G. Town, "Magnetically coupled high-gain Y-source isolated DC/DC converter", *IET Power Electron.*, vol. 7, no. 11, pp. 2817–2824, Nov. 2014.

168. J. Chen, B. Hwang, C. Kung, W. Tai, Y. Hwang, "A New Single-Inductor Quadratic Buck Converter using Average-Current-Mode Control without Slope-Compensation", Conf. Proc, IEEE 5th Conf. on Industrial Electronics and Applications, IEEE-ICIEA'10, Taichung (Taiwan), pp. 1082–1087, June 15–17, 2010.

169. N. Kondrath, M. Kazimierczuk, "Analysis and design of common-diode tapped inductor PWM buck converter", Conf. Proc, Electrical Manufacturing and Coil Winding Conf., Nashville (TN), September 29–30, 2009.

170. K. Yao, M. Ye, M. Xu, F. Lee, "Tapped-inductor buck converter for high-step-down DC-DC conversion", *IEEE Trans. Power Electron.*, vol. 20, no. 4, pp. 775–780, July 2005.

171. K. Nishijima, K. Abe, D. Ishida, T. Nakano, T. Nabeshima, T. Sato, K. Harada, "A Novel Tapped-Inductor Buck Converter for Divided Power Distribution System", Conf. Proc, IEEE 37th Power Electronics Specialists Conf., IEEE-PESC'06, Jeju (South Korea), pp. 1–6, June 18–22, 2006.

172. L. Weichen, X. Xin, L. Chushan, L. Wuhua, H. Xiangning, "Interleaved high step-up ZVT converter with built-in voltage doubler cell for distributed PV generation system", *IEEE Trans. Power Electron.*, vol. 28, no. 1, pp. 300–313, Jan. 2013.

173. D. Wang, X. He, R.. Zhao, "ZVT interleaved boost converters with built-in voltage doubler and current auto-balance characteristic", *IEEE Trans. Power Electron.*, vol. 23, no. 6, pp. 2847–2854, Nov. 2008.

174. W. Li, Y. Zhao, Y. Deng, X. He, "Interleaved converter with voltage multiplier cell for high step-up and high-efficiency conversion", *IEEE Trans. Power Electron.*, vol. 25, no. 9, pp. 2397–2408, Sep. 2010.

175. W. Li, W. Li, M. Ma, Y. Deng, X. He, "A non-isolated high step-up converter with built-in transformer derived from its isolated counterpart", Conf. Proc, IEEE 36th Annual Conf. on Industrial Electronics Society, IEEE-IECON'10, Glendale (USA), pp. 3173–3178, November 7–10, 2010.

176. W. Li, X. He, "An interleaved winding-coupled boost converter with passive lossless clamp circuits", *IEEE Trans. Power Electron.*, vol. 22, no. 4, pp. 1499–1507, July 2007.

177. K. Yao, Y. Meng, F. Lee, "A novel winding coupled-buck converter for high-frequency, high step-down DC/DC conversion", *IEEE Trans. Power Electron.*, vol. 20, no. 5, pp. 1017–1024, Sep. 2005.

178. S. Tseng, C. Hsu, "Interleaved step-up converter with a single capacitor snubber for PV energy conversion applications", *Int. J. Electr. Power Energy Syst.*, vol. 53, pp. 909–922, Dec. 2013.

179. O. Pelan, N. Muntean, O. Cornea, "High Voltage Conversion Ratio, Switched C & L Cells, Step-Down DC-DC Converter", Conf. Proc, IEEE Energy Conversion Congress and Exposition, IEEE-ECCE'13, Denver (USA), pp. 5580–5585, September 15–19, 2013.

180. S. Wibowo, Z. Ting, M. Kono, T. Taura, Y. Kobori, H. Kobayashi, "Analysis of Coupled Inductors for Low-Ripple Fast-Response Buck Converter", Conf. Proc, IEEE Asia Pacific Conf. on Circuits and Systems, IEEE-APCCAS'08, Macao (China), pp. 1860–1863, November 30–December 3, 2008.

181. G. Zhu, B. A. McDonald, K. Wang, "Modeling and analysis of coupled inductors in power converters", *IEEE Trans. Power Electron.*, vol. 26, no. 5, pp. 1355–1363, May 2011.

182. T. Schmid and A. Ikriannikov, "Magnetically Coupled Buck Converters", Conf. Proc, IEEE Energy Conversion Congress and Exposition, IEEE-ECCE'13, Denver (USA), pp. 4948–4954, September 15–19, 2013.

183. L. F. Costa, S. A. Mussa, I. Barbi, "Multilevel Buck DC-DC Converter for High Voltage Application", Conf. Proc, IEEE 10th Intl. Conf. on Ind. Applications, IEEE-INDUSCON'12, Fortaleza (Brazil), pp. 1–8, November 5–7, 2012.

184. J. Balestero, F. Tofoli, G. Torrico-Bascop´e, F. Seixas, "A DC–DC Converter Based on the Three-State Switching Cell for High Current and Voltage Step-Down Applications", *IEEE Trans. Power Electron.*, vol. 28, no. 1, pp. 398–407, Jan. 2013.

185. F. Luo, H. Ye, *"Advanced DC-DC Converters"*, 2nd Edition, UK: CRC Press, https://www.routledge.com/Advanced-DCDC-Converters/Luo-Ye/p/book/9781498774901, Dec 2016.

186. F. Luo and Ye H., "DC/DC Conversion Techniques and Nine Series Luo-Converters", *M. H. Rashid-Power electronics handbook Devices, Circuits and Applications*, United States: Elsevier, 2011.

187. Luo, F. L., "Re-lift converter: Design, rest, simulation and stability analysis", *IEE Proc. Electric Power Appl., IET*, vol. 145, no. 4, pp. 315–325, July 1998.

188. Luo, F. L. "Negative output Luo-Converters: Voltage lift technique", *IEE Proc. Electric Power Appl., IET*, vol. 146, no. 2, pp. 208–224, March 1999.

189. L. Fang Lin, "Six self-lift DC-DC converters, voltage lift technique", *IEEE Trans. Ind. Electron.*, vol. 48, no. 6, pp. 1268–1272, Dec. 2001.

190. F. Z. Peng, "Z-source inverter", *IEEE Trans. Ind. Appl.*, vol. 39, no. 2, pp. 504–510, March 2003.

191. H. Abu-Rub, M. Malinowski, K. Al-Haddad, *"Power Electronics for Renewable Energy Systems, Transportation and Industrial Applications"*, United States: John Wiley & Sons, Ltd. 2014.

192. B. Ge, H. Abu-Rub, F. Z. Peng, Q. Lei, A. de Almeida, F. Ferreira, D. Sun, Y. Liu "An energy-stored quasi-Z-source inverter for application to photovoltaic power system", *IEEE Trans. Ind. Electron.*, vol. 60, no. 10, pp. 4468–4481, Oct. 2013.

193. Y. P. Siwakoti, F. Z. Peng, F. Blaabjerg, P. C. Loh, G. E. Town, "Impedance-source networks for electric power conversion Part I: A topological teview", *IEEE Trans. Power Electron.*, vol. 30, no. 2, pp. 699–716, Feb. 2015.
194. O. Ellabban, H. Abu-Rub, "Z-source inverter: Topology improvements review", *IEEE Ind. Electron. Mag.*, vol. 10, no. 1, pp. 6–24, March 2016.
195. H. Shen, B. Zhang, D. Qiu, "Hybrid Z-source boost DC–DC converters", *IEEE Trans. Ind. Electron.*, vol. 64, no. 1, pp. 310–319, Jan. 2017.
196. A. Torkan, M. Ehsani, "A novel nonisolated Z-source DC–DC converter for photovoltaic applications", *IEEE Trans. Ind. Appl.*, vol. 54, no. 5, pp. 4574–4583, Sep. 2018.
197. M. Meraj, M. S. Bhaskar, A. Iqbal, N. Al-Emadi, S. Rahman, "Interleaved multilevel boost converter with minimal voltage multiplier components for high-voltage step-up applications", *IEEE Trans. Power Electron.*, vol. 35, no. 12, pp. 12816–12833, Dec. 2020.
198. C. Young, M. Chen, T. Chang, C. Ko, K. Jen, "Cascade Cockcroft-Walton voltage multiplier applied to transformer less high step-Up DC-DC converter", *IEEE Trans. Ind. Electronics*, vol. 60, no. 2, pp. 523–537, Feb. 2013.
199. L. Muller, J. Kimball, "High gain DC–DC converter based on the Cockcroft–Walton multiplier", *IEEE Trans. Power Electron.*, vol. 31, no. 9, pp. 6405–6415, Sep. 2016.
200. J. Wang, S. W. H. de Haan, J. A. Ferreira, P. Luerkens, "Complete Model of Parasitic Capacitances in a Cascade Voltage Multiplier in the High Voltage Generator", Conf. Proc., IEEE ECCE Asia Downunder, Melbourne (Australia), pp. 18–24, June 3–6, 2013.
201. W. Bin, K. Smedley, "A family of two-switch boosting switched capacitor converters", *IEEE Trans. Power Electron*, vol. 30, no. 10, pp. 5413–5424, Oct. 2015.
202. Y. J. A. Alcazar, D. de Souza Oliveira, F. L. Tofoli, R. P. Torrico-Bascope, "DC-DC non isolated boost converter based on the three-state switching cell and voltage multiplier cells", *IEEE Trans. Ind. Electron*, vol. 60, no. 10, pp. 4438–4449, Oct. 2013.
203. J. Rosas-Caro, J. Mayo-Maldonado, R. Cabrera, A. Rodriguez, S. Nacu, R. Castillo-Ibarra, "A family of DC-DC multiplier converters", *Engineering Letters, International Association of Engineers (IAENG)*, 10 Feb. 2011.
204. M. S. Bhaskar, R. Al-ammari, M. Meraj, A. Iqbal, S. Padmanaban, "Modified multilevel buck–boost converter with equal voltage across each capacitor: Analysis and experimental investigations", *IET Power Electron.*, vol. 12, no. 13, pp. 3318–3330, 2019.
205. P. K. Maroti, M. S. B. Ranjana, D. K. Prabhakar, "A novel high gain switched inductor multilevel buck-boost DC-DC converter for solar applications", Conf. Proc., IEEE 2nd Intl. Conf. on Electrical Energy Systems, IEEE-ICEES'14, Chennai (India), pp. 152–156, January 7–9, 2014.
206. M. S. Bhaskar, N. S. Reddy, R. Kumar, "Non-isolated dual output hybrid DC-DC multilevel converter for photovoltaic applications", 2nd Intl. Conf. on Devices, Circuits and Systems (ICDCS), pp. 1–6, Coimbatore, India, March 2014.
207. J. C. Rosas-Caro, J. Mayo-Maldonado, J. Valdez, R. Salas-Cabrera, A. Rodriguez, E. Salas-Cabrera, H. Cisneros-Villegas, J. Gonzalez-Hernandez, "Multiplier SEPIC converter", Conf. Proc, 21st Intl. Conf. on Electrical Communications and Computers, CONIELECOMP'11, Cholula, (Mexico), February 28–March 2, 2011.
208. J. C. Rosas-Caro, V. Sanchez, R. Vazquez-Bautista, L. Morales-Mendoza, J. Mayo-Maldonado, P. Garcia-Vite, R. Barbosa, "A novel DC-DC multilevel SEPIC converter for PEMFC systems", *Int. J. Hydrogen Energy*, vol. 41, no. 48, pp. 23401–23408, Dec. 2016.
209. M. S. B. Ranjana, N. S. Reddy, R. Kumar, "A novel SEPIC based dual output DC-DC converter for solar applications", Conf. Proc., IEEE Power and Energy Systems Conf.: Towards Sustain. Energy, Bangalore (India), pp. 1–5, March 13–15, 2014.
210. M. Mousa, M. Ahmed, M. Orabi, "A Switched Inductor Multilevel Boost converter", Conf. Proc., IEEE Intl. Conf. on Power and Energy, IEEE-PECon'10, Kuala Lumpur (Malaysia), pp. 819–823, November 29–December 1, 2010.

211. M. S. B. Ranjana, N. S. Reddy, R. Kumar, "A novel Non-Isolated Switched Inductor Floating Output DC-DC Multilevel Boost Converter for Fuel Cell Applications", Conf. Proc., IEEE Students' Conf. on Electrical, Electronics and Computer Science, IEEE-SCEECS'14, Bhopal (India), pp. 1–5, March 1–2, 2014.

212. S. B. Mahajan, P. Sanjeevikumar, P. Wheeler, F. Blaabjerg, M. Rivera, R. Kulkarni, "X-Y Converter Family: A New Breed of Buck Boost Converter for High Step-Up Renewable Energy Applications", Conf. Proc., IEEE Intl. Conf. on Automatica IEEE-ICA-ACCA'16, Curico (Chile), pp. 1–8, October 19–21, 2016.

213. S. B. Mahajan, P. Sanjeevikumar, F. Blaabjerg, Rishi Kulkarni, Shridhar Seshagiri, Amin Hajizadeh, "Novel LY Converter Topologies for High Gain Transfer Ratio—A New Breed of XY Family", Conf. Proc., IET 4th IET Intl. Conf. On Clean Energy and Technology, IET-CEAT'16, Kuala Lumpur (Malaysia), pp. 1–8, November 14–15, 2016.

214. M. S. Bhaskar, P. Sanjeevikumar, A. Iqbal, M. Meraj, A. Howeldar, J. Kamuruzzaman, "L-L Converter for Fuel Cell Vehicular Power Train Applications: Hardware Implementation of Primary Member of X-Y Converter Family", IEEE International Conference on Power Electronics, Drives and Energy Systems (PEDES), pp. 1–6, Chennai, India. December 18–21, 2018.

215. M. S. Bhaskar, P. Sanjeevikumar, J. B. Holm-Nielsen, J. K. Pedersen, Z. Leonowicz, "2L-2L Converter: Switched Inductor Based High Voltage Step-up Converter for Fuel Cell Vehicular Applications", IEEE International Conference on Environment and Electrical Engineering, and IEEE Industrial and Commercial Power Systems Europe, pp. 1–6, Genova, Italy, June 11–14, 2019.

216. M. S. Bhaskar, M. Meraj, A. Iqbal, R. Al-ammari, S. Padmanaban, "New DC-DC Multilevel Configurations of 2L-Y Boost Converters with High Voltage Conversion Ratio for Renewable Energy Applications", IEEE 28th Intl. Symposium on Industrial Electronics (ISIE), pp. 2527–2532, Vancouver, Canada, June 2–14, 2019.

217. M. S. Bhaskar, P. Sanjeevikumar, F. Blaabjerg, J. B. Holm-Nielsen, D. M. Ionel, "L-L and L-2L Multilevel Boost Converter Topologies with Voltage Multiplier with L-L and L-2L Converter of XY Family", IEEE 59th International Scientific Conference on Power and Electrical Engineering of Riga Technical University (RTUCON), pp. 1–6, Riga, Latvia, November 12–13, 2018.

218. P. K. Maroti, P. Sanjeevikumar, M. Bhaskar, F. Blaabjerg, V. Ramachandaramurthy, P. Siano, V. Fedak, "A Novel 2L-Y DC-DC Converter Topologies for High Conversion ratio Renewable Application", IEEE Conference on Energy Conversion (CENCON), Kuala Lumpur, Malaysia, pp. 323–328, October 30–31, 2017.

219. S. B. Mahajan, P. Sanjeevikumar, K. M. Pandav, V. Fedák, F. Blaabjerg, V. Ramachandramurthy, "New 2LC-Y DC-DC Converter Topologies for High-Voltage /Low-Current Renewable Application: New Members of X-Y Converter Family", Conf. Proc. of 19th IEEE Intl. Conf. on Electrical Drives and Power Electronics, IEEE-EDPE'17, Dubrovnik, Croatia (Europe), pp. 139–146, October 4–6, 2017.

220. S. Sadaf, N. A. Al-Emadi, A. Iqbal, M. S. Bhaskar, M. Meraj, "New High Gain 2LC-Y Multilevel-Boost-Converter (2LC-Y MBC) Topologies for Renewable Energy Conversion: Members of X-Y Converter Family", IEEE 28th Intl. Symposium on Industrial Electronics (ISIE), pp. 2647–2652, Vancouver, Canada, June 2–14, 2019.

221. M. S. Bhaskar, S. Padmanaban, P. Wheeler, F. Blaabjerg, P. Siano, "A New Voltage Doubler Based DC-DC $2LC_m$-Y Power Converter Topologies for High-Voltage/Low-Current Renewable Energy Applications", IEEE Transportation Electrification Conference and Expo (ITEC), pp. 1–6, Long Beach, CA, USA, June 13–15, 2018.

222. S. B. Mahajan, P. Sanjeevikumar, V. Fedák, F. Blaabjerg, P. Wheeler, V. Ramachandramurthy, "L-L Multilevel Boost Converter Topology For Renewable Energy Applications: A New Series Voltage Multiplier L-L Converter of XY Family",

Conf. Proc. of 19th IEEE Intl. Conf. on Electrical Drives and Power Electronics, IEEE-EDPE'17, Dubrovnik, Croatia (Europe), pp. 133–138, October 4–6, 2017.

223. P. Sanjeevikumar, M. S. Bhaskar, F. Blaabjerg, Y. Yang, "A New DC-DC Multilevel Breed of XY Converter Family for Renewable Energy Applications: LY Multilevel Structured Boost Converter", 44th Annual Conference of the IEEE Industrial Electronics Society, pp. 6110–6115, Washington, DC, USA, Oct. 2018.

224. F. Zhang, L. Du, F. Peng, Z. Qian, "A New Design Method for High Efficiency DC-DC Converters with Flying Capacitor Technology", Twenty-First Annual IEEE Applied Power Electron. Conf. and Expo., pp. 1–5, Dallas, TX, USA, March 19–23, 2006.

225. F. Zhang, F. Peng, Z. Qian, "Study of the Multilevel Converters in DC-DC Applications", IEEE 35th Annual Power Electron. Spec. Conf. (IEEE Cat. No. 04CH37551), pp. 1702–1706, Aachen, Germany, June 20–25, 2004.

226. Z. Pan, F. Zhang, F. Z. Peng, "Power Losses and Efficiency Analysis of Multilevel DC-DC Converters", Twentieth Annual IEEE Applied Power Electron. Conf. and Expo., pp. 1393–1398, Austin, TX, USA, March 6–10, 2005.

227. A. Viraj, G. Amaratunga, "Analysis of Switched Capacitor DC-DC Step Down Converter", IEEE Intl. Symp. on Circuits and Systems (IEEE Cat. No. 04CH37512), pp. 836–839, Vancouver, BC, Canada, May 23–26, 2004.

228. W. Harris, K. Ngo, "Power switched-capacitor DC-DC converter: Analysis and design", *IEEE Trans. Aero. Electron. Syst.*, vol. 33, no. 2, pp. 386–395, April 1997.

229. M. Makowski, D. Maksimovic, "Performance limits of switched-capacitor DC-DC converters", Proceed. of PESC '95 - Power Electron. Specialist Conf., pp. 1215–1221, Atlanta, GA, USA, June 18–22, 1995.

230. K. Eguchi, R. Rubpongse, A. Shibata, T. Ishibashi, "Synthesis and analysis of a cross-connected Fibonacci dc/dc converter with high voltage gain", *Energy Rep.*, vol. 6, pp. 130–136, Feb. 2020.

231. A. Kushnerov, S..Yaakov, "Algebraic Synthesis of Fibonacci Switched Capacitor Converters", IEEE International Conference on Microwaves, Communications, Antennas and Electronic Systems (COMCAS 2011), pp. 1–4, Tel Aviv, Israel, November 7–9, 2011.

232. F. Z. Peng, F. Zhang, Z. Qian, "A magnetic-less DC-DC converter for dual-voltage automotive systems", *IEEE Trans. Ind. Appl.*, vol. 39, no. 2, pp. 511–518, March 2003.

233. O. Mak, Y. Wong, A. Ioinovici, "Step-up DC power supply based on a switched-capacitor circuit", *IEEE Trans. Ind. Electron.*, vol. 42, no. 1, pp. 90–97, Feb. 1995.

234. T. Umeno, K. Takahashi, I. Oota, F. Ueno, T. Inoue, "New switched-capacitor DC-DC converter with low input current ripple and its hybridization", 33rd Midwest Symposium on Circuits and Systems, pp. 1091–1094, Calgary, Alberta, Canada, August 12–14, 1990.

235. Khan, F. H. *Modular DC-DC Converters*. Ph.D. Thesis, The University of Tennessee, Knoxville, TN, USA, May 2007.

236. P. Maroti, P. Sanjeevikumar, J. Holm-Nielsen, M. S. Bhaskar, M. Meraj, A. Iqbal, "A new structure of high voltage gain SEPIC converter for renewable energy applications", *IEEE Access*, vol. 7, pp. 89857–89868, 2019.

237. E. Ozsoy, P. Sanjeevikumar, F. Blaabjerg, D. Ionel, U. Kalla, M. Bhaskar, "Control of High Gain Modified SEPIC Converter: A Constant Switching Frequency Modulation Sliding Mode Controlling Technique", IEEE Intl. Power Electron. and Appl. Conf. and Exposition (PEAC), pp. 1–6, Shenzhen, China, November 4–7, 2018.

238. M. S. Bhaskar, P. Sanjeevikumar, J. Pedersen, J. Holm-Nielsen, Z. Leonowicz, "XL Converters—New Series of High Gain DC-DC Converters for Renewable Energy Conversion", IEEE Intl. Conf. on Environment and Electrical Engg. and IEEE Ind. and Commercial Power Systems Europe, pp. 1–6, Genova, Italy, June 11–14, 2019.

239. M. Alghaythi, R. O'Connell, N. Islam, M. Khan, J. M. Guerrero, "A high step-up interleaved DC-DC converter with voltage multiplier and coupled inductors for renewable energy systems", *IEEE Access*, vol. 8, pp. 123165–123174, 2020.

240. X. Fan, H. Sun, Z. Yuan, Z. Li, R. Shi, N. Ghadimi, "High voltage gain DC/DC converter using coupled inductor and VM techniques", *IEEE Access*, vol. 8, pp. 131975–131987, 2020.

241. W. Li, X. He, "A bick interleaved DC–DC converters deduced from a basic cell with winding-cross-coupled inductors (WCCIs) for high step-up or step-down conversions", *IEEE Trans. Power Electron.*, vol. 23, no. 4, pp. 1791–1801, July 2008, doi: 10/fwxgdt.

242. J. Ai, M. Lin, M. Yin, "A family of high step-up cascade DC–DC converters with clamped circuits", *IEEE Trans. Power Electron.*, vol. 35, no. 5, pp. 4819–4834, May 2020.

243. F. L. Luo, H. Ye, "Positive output cascade boost converters", *IEE Proc. - Electric Power Appl.*, vol. 151, no. 5, pp. 590–606, Sep. 2004.

244. C. Young, M. Chen, T. Chang, C. Ko, K. Jen, "Cascade Cockcroft–Walton voltage multiplier applied to transformerless high step-up DC–DC converter", *IEEE Trans. Ind. Electron.*, vol. 60, no. 2, pp. 523–537, Feb. 2013.

245. Y. Berkovich, B. Axelrod, A. Shenkman, "A novel diode-capacitor voltage multiplier for increasing the voltage of photovoltaic cells", 11th Workshop on Control and Modeling for Power Electronics, pp. 1–5, Zurich, Switzerland, August 17–20, 2008.

246. A. Tomaszuk, A. Krupa, "High efficiency high step-up DC/DC converters: A review", *Bull. Polish Acad. Sci. Technol. Sci.*, vol. 59, no. 4, pp. 475–483, 2011.

247. L. Schmitz, D. C. Martins, R. F. Coelho, "Comprehensive conception of high step-up DC–DC converters with coupled inductor and voltage multipliers techniques", *IEEE Trans. Circuits Syst. I: Regular Papers*, vol. 67, no. 6, pp. 2140–2151, June 2020.

248. N. Priyadarshi, P. Sanjeevikumar, J. Holm-Nielsen, M. S. Bhaskar, F. Azam, "Internet of things augmented a novel PSO-employed modified zeta converter-based photovoltaic maximum power tracking system: Hardware realisation", *IET Power Electron.*, Jan. 2020 (early access).

249. M. S. Bhaskar, N. S. Reddy, R. Kumar, "A novel high gain floating output DC-DC multilevel boost converter for fuelcell applications", Conf. Proc., IEEE Conf. on Circuits, Power and Computing Technologies, IEEE-ICCPCT'14, Nagarcoil (India), pp. 291–295, March 20–21, 2014.

250. E. Babaei, H. Mashinchi Maheri, M. Sabahi, S. H. Hosseini, "Extendable nonisolated high gain DC-DC converter based on active–passive inductor cells", *IEEE Trans. Ind. Electron.*, vol. 65, no. 12, pp. 9478–9487, Dec. 2018.

251. M. S. Bhaskar, P. Sanjeevikumar, F. Blaabjerg, J. Holm-Nielsen, M. Mitolo, "Chain of X-Y Power Novel DC-DC Converters with Synchronous Grounded Switching for High Step-Up Renewable Power Applications", IEEE Intl. Conf. on Environment and Electrical Engg. and IEEE Ind. and Commercial Power Systems Europe, pp. 1–6, Madrid, Spain, June 9–12, 2020.

252. M. Lakshmi, S. Hemamalini, "Nonisolated high gain DC–DC converter for DC microgrids", *IEEE Trans. Ind. Electron.*, vol. 65, no. 2, pp. 1205–1212, Feb. 2018.

253. M. Muhammad, M. Armstrong, M. Elgendy, "A nonisolated interleaved boost converter for high-voltage gain applications", *IEEE J. Emerg. Select. Top. Power Electron.*, vol. 4, no. 2, pp. 352–362, June 2016.

254. A. Alzahrani, M. Ferdowsi, P. Shamsi, "A family of scalable non-isolated interleaved DC-DC boost converters with voltage multiplier cells", *IEEE Access*, vol. 7, pp. 11707–11721, 2019.

255. V. A. K. Prabhala, P. Fajri, V. S. P. Gouribhatla, B. P. Baddipadiga, M. Ferdowsi, "A DC–DC converter with high voltage gain and two input boost stages", *IEEE Trans. Power Electron.*, vol. 31, no. 6, pp. 4206–4215, June 2016.

256. C. A. Soriano-Rangel, J. C. Rosas-Caro, F. Mancilla-David, "An optimized switching strategy for a ripple-canceling boost converter", *IEEE Trans. Ind. Electron.*, vol. 62, no. 7, pp. 4226–4230, July 2015.

257. M. Bhaskar, R. Kulkarni, P. Sanjeevikumar, P. Siano, F. Blaabjerg, "Hybrid non-isolated and non inverting Nx interleaved DC-DC multilevel boost converter for renewable energy applications", Conf. Proc., IEEE 16th Intl. Conf. on Environment and Electrical Engg., pp. 1–6, Florence (Italy), June 7–10, 2016.

258. M. Bhaskar, D. Almakhles, P. Sanjeevikumar, F. Blaabjerg, U. Subramaniam, D. M. Ionel, "Analysis and Investigation of Hybrid DC–DC Non-Isolated and Non-Inverting Nx Interleaved Multilevel Boost Converter (Nx-IMBC) for High Voltage Step-Up Applications: Hardware Implementation", *IEEE Access*, vol. 8, pp. 87309–87328, 2020.

259. C. Pan, C.-F. Chuang, C.-C. Chu, "A novel transformer-less adaptable voltage quadrupler DC converter with low switch voltage stress", *IEEE Trans. Power Electron.*, vol. 29, no. 9, pp. 4787–4796, Sep. 2014.

260. P. Sanjeevikumar, G. Grandi, F. Blaabjerg, P. Wheeler, P. Siano, M. Hammami, "A comprehensive analysis and hardware implementation of control strategies for high output voltage DC-DC boost power converter", *Int. J. Comp. Intell. Syst.*, vol. 10, no. 1, pp. 140–152, Jan. 2017.

261. M. S. Bhaskar, P. Sanjeevikumar, O. Ojo, M. Rivera, R. M. Kulkarni, "Non-isolated and inverting Nx multilevel boost converter for photovoltaic DC link applications", Conf. Proc., IEEE Intl. Conf. on Automatica IEEE-ICA-ACCA'16, Curico (Chile), pp. 1–8, October 19–21, 2016.

Index